AS/A-LEVEL

STUDENT GUIDE

AQA

Geography

Component 1: Physical geography

Water and carbon cycles

Hot desert systems and landscapes

Coastal systems and landscapes

Glacial systems and landscapes

David Redfern

HODDER
EDUCATION
AN HACHETTE UK COMPANY

Hodder Education, an Hachette UK company, Blenheim Court, George Street, Banbury, Oxfordshire OX16 5BH

Orders

Bookpoint Ltd, 130 Park Drive, Milton Park, Abingdon, Oxfordshire OX14 4SB

tel: 01235 827827

fax: 01235 400401

e-mail: education@bookpoint.co.uk

Lines are open 9.00 a.m.–5.00 p.m., Monday to Saturday, with a 24-hour message answering service. You can also order through the Hodder Education website: www.hoddereducation.co.uk

© David Redfern 2016

ISBN 978-1-4718-6404-9

Impression number 5 4 3 2 1

Year 2020 2019 2018 2017 2016

Cover photo: pure-life-pictures/Fotolia. Other photos: David Redfern (p. 66), nauke13/Fotolia (p. 72).

Typeset by Integra Software Services Pvt Ltd, Pondicherry, India

Printed in Italy

Hachette UK's policy is to use papers that are natural, renewable and recyclable products and made from wood grown in sustainable forests. The logging and manufacturing processes are expected to conform to the environmental regulations of the country of origin.

Contents

Content Guidance

Questions & Answers

■ Getting the most from this book

Exam tips

Advice on key points in the text to help you learn and recall content, avoid pitfalls, and polish your exam technique in order to boost your grade.

Knowledge check

Rapid-fire questions throughout the Content Guidance section to check your understanding.

Knowledge check answers

1 Turn to the back of the book for the Knowledge check answers.

Summaries

■ Each core topic is rounded off by a bullet-list summary for quick-check reference of what you need to know.

Exam-style questions

Commentary on the questions

Tips on what you need to do to gain full marks, indicated by the icon **e**

Sample student answers

Practise the questions, then look at the student answers that follow.

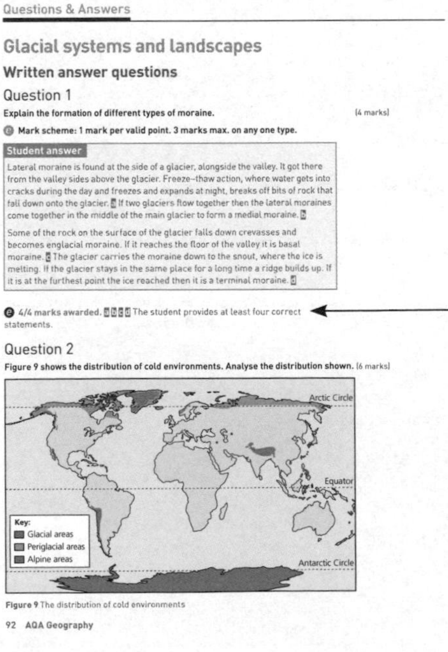

Commentary on sample student answers

Read the comments (preceded by the icon **e**) showing how many marks each answer would be awarded in the exam and exactly where marks are gained or lost.

Glacial systems and landscapes

Written answer questions

Question 1

Explain the formation of different types of moraine. (4 marks)

e Mark scheme: 1 mark per valid point. 3 marks max. on any one type.

Student answer

Lateral moraine is found at the side of a glacier, alongside the valley. It got there from the valley sides above the glacier. Freeze–thaw action, where water gets into cracks during the day and freezes and expands at night, breaks off bits of rock that fall down onto the glacier. If two glaciers flow together then the lateral moraines come together in the middle of the main glacier to form a medial moraine.

Some of the rock on the surface of the glacier falls down crevasses and becomes englacial moraine. If it reaches the floor of the valley it is basal moraine. The glacier carries the moraine down to the snout, where the ice is melting. If the glacier stays in the same place for a long time a ridge builds up. If it is at the furthest point the ice reached then it is a terminal moraine.

e 4/4 marks awarded. The student provides at least four correct statements.

Question 2

Figure 9 shows the distribution of cold environments. Analyse the distribution shown. (6 marks)

Key:
■ Glacial areas
■ Periglacial areas
■ Alpine areas

Figure 9 The distribution of cold environments

■About this book

Much of the knowledge and understanding needed for AS and A-level geography builds on what you have learned for GCSE geography, but with an added focus on key geographical concepts and depth of knowledge and understanding of content. This guide offers advice for the effective revision of **Physical geography**, which all students need to complete.

The first part of the A-level Paper 1 external exam paper tests your knowledge and application of aspects of Physical geography with a particular focus on the Water and carbon cycles, and *one* of Hot desert systems and landscapes, Coastal systems and landscapes and Glacial systems and landscapes. The whole exam lasts 2 hours 30 minutes and the unit makes up 40% of the A-level award. Three of the same topics (all *except* Hot desert systems and landscapes) make up 50% of the AS Paper 1 (lasting 1 hour and 30 minutes) but only *one* of them needs to be studied. More information on the exam papers is given in the Questions & Answers section (pages 55–58).

To be successful in this unit you have to understand:
- the key ideas of the content
- the nature of the assessment material — by reviewing and practising sample structured questions
- how to achieve a high level of performance within them

This guide has two sections:

Content Guidance — this summarises some of the key information that you need to know to be able to answer the examination questions with a high degree of accuracy and depth. In particular, the meaning of keys terms is made clear and some attention is paid to providing details of case study material to help to meet the spatial context requirement within the specification.

Questions & Answers — this includes some sample questions similar in style to those you might expect in the exam. There are some sample student responses to these questions as well as detailed analysis, which will give further guidance in relation to what exam markers are looking for to award top marks.

The best way to use this book is to read through the relevant topic area first before practising the questions. Only refer to the answers and examiner comments after you have attempted the questions.

Content Guidance

This section outlines the following areas of the AQA AS and A-level Geography specifications:

■ Water and carbon cycles
■ Hot desert systems and landscapes
■ Coastal systems and landscapes
■ Glacial systems and landscapes

■ Water and carbon cycles

Water and carbon cycles as natural systems

In geography, two general types of system are recognised:

1 **A closed system:** where there is transfer of energy, but not matter, between the system and its surroundings. Planet Earth is an example of such a system.

2 **An open system:** where a system receives inputs and transfers outputs of energy or matter across the boundaries within it and with its surroundings. Most natural systems, such as the water cycle and carbon cycle, are open systems.

Open systems have common features that are important elements in their own right:

■ **Inputs** are those elements that enter a system to be processed. They are fed into the system in order to create outputs.

■ **Outputs** are the outcome of processing within the system. Outputs may be of use to the next element in the system, or they may be unintended outcomes that may not be of use.

■ **Stores** (or **components**) are where amounts of energy or matter are held, and not transferred until the appropriate processes are in place to move them.

■ **Transfers** or flows involve the movement of energy or matter through the system and enable inputs to become outputs. These may involve processes that create change.

Open systems tend to adjust themselves to flows of energy and/or matter by modifying the interrelationships between different elements of the system, so that input and output flows balance each other out, resulting in a steady state for the system, known as dynamic equilibrium. This kind of adjustment is called self-regulation, and much of physical geography can be understood in part as the study of self-regulating systems.

Feedback is probably one of the most important aspects of systems theory. It occurs when one element of a system changes because of an outside influence. This will upset the dynamic equilibrium, or state of balance, and affect other components in the system. **Negative feedback** is when a system acts by *lessening* the effect of the original change and ultimately reversing it. **Positive feedback** occurs within a system where a change causes a further, or snowball, effect, *continuing* or even accelerating the original change.

The water cycle

The world's water is always in the process of movement — the natural water cycle describes the continuous movement of water on, above and below the surface of the Earth. Water is always changing states between liquid, vapour and ice, with some of these processes happening within seconds and others over much longer periods of time.

The distribution and size of water stores

96% of the world's water is saline seawater; of the total freshwater, over 68% is locked up in ice and glaciers. Another 30% of freshwater is in the ground. Fresh surface-water sources, such as rivers and lakes, constitute only about 1/150th of 1% of total water. Yet, rivers and lakes are the sources of most of the freshwater people use every day.

The hydrosphere (oceans)

Much more water is 'in storage' for long periods of time than is actually moving through the water cycle. The oceans hold the vast majority of all water on Earth. It is also estimated that the oceans supply about 90% of the evaporated water that goes into the water cycle.

Although over the short term of hundreds of years the oceans' volumes do not change much, the amount of water in the oceans does change over the long term. During the last ice age, sea levels were lower by as much as 120 m, which allowed humans to cross to North America from Asia at the (now underwater) Bering Strait. Similarly, the English Channel was dry. During such colder climatic periods more ice caps and glaciers form, and enough of the global water supply accumulates as ice to lower sea levels. The reverse is true during warm periods. During the last major global 'warm spell', 125,000 years ago, the seas were about 5.5 m higher than they are now.

The atmosphere

Although the atmosphere may not be a great store of water (only about 0.001% of the total Earth's water), it is the main vector that moves water around the globe. Evaporation and transpiration change liquid water into vapour, which then ascends into the atmosphere due to rising air currents. Cooler temperatures at altitude allow the vapour to condense into clouds and strong winds move the clouds around the world until the water falls as precipitation to replenish the land-based parts of the water cycle.

About 90% of water in the atmosphere is produced by evaporation from water bodies, while the other 10% comes from transpiration from plants. A very small amount of water vapour enters the atmosphere through sublimation, the process by which water changes from a solid (ice or snow) to a gas, bypassing the liquid phase. Clouds are the

> **Exam tip**
>
> Systems theories, involving various forms of feedback, feature several times in this book. Make sure you understand these concepts fully.

> **Exam tip**
>
> It is important that proportions, or percentages, of relative amounts of water are learnt and understood in this section.

most visible manifestation of atmospheric water, but even clear air contains water — water in particles that are too small to be seen.

The cryosphere (ice caps, ice shelves, sea ice and glaciers)

The vast majority — almost 90% — of Earth's ice mass is in Antarctica, while the Greenland ice cap contains 10% of the total global ice mass. Collectively ice caps and glaciers cover about 10% of the Earth's surface. An ice shelf is a floating extension of land ice. Ice shelves in Antarctica cover more than 1.6 million km² (an area the size of Greenland), fringing 75% of the continent's coastline, and covering 11% of its total area.

Sea ice is frozen ocean water, surrounding several polar regions of the world. On average, sea ice covers up to 25 million km², an area 2.5 times the size of Canada. One difference between sea ice and ice shelves is that sea ice is free-floating; the sea freezes and unfreezes each year, whereas ice shelves are firmly attached to the land.

Lithosere (land-based) storage

Freshwater storage

Surface freshwater includes water courses of all sizes, from large rivers to small streams, ponds, lakes, reservoirs and freshwater wetlands. Freshwater represents only about 2.5% of all water on Earth and freshwater lakes and swamps account for a mere 0.29% of the Earth's freshwater. A total of 20% of all surface freshwater is in one lake — Lake Baikal in Asia. Another 20% is stored in the Great Lakes of North America. Rivers hold only about 0.006% of total freshwater reserves.

Groundwater storage

Some of the precipitation that falls onto the land infiltrates deep into the ground to become groundwater. Large quantities of water are held deep underground in zones called aquifers. Water from these aquifers can take thousands of years to move back into the surface environment, if at all.

Factors driving change in the water cycle

Evaporation

Evaporation is the process by which water changes from a liquid to a gas or vapour. Heat (energy) is necessary for evaporation to occur. Energy is used to break the bonds that hold water molecules together, which is why water easily evaporates at the boiling point (100°C) and evaporates much more slowly at the freezing point. Evaporation from the oceans is the primary mechanism supporting the surface-to-atmosphere portion of the water cycle.

Evapotranspiration

Evapotranspiration refers to the combination of evaporation and transpiration. It is defined as the water lost to the atmosphere from the ground surface, evaporation from the capillary fringe of the water table and the transpiration of groundwater by plants whose roots tap the capillary fringe of the water table. The transpiration aspect of evapotranspiration is the process by which water is lost from a plant through the stomata in its leaves.

Exam tip

It is likely that questions will make use of the terms lithosphere, hydrosphere, cryosphere and atmosphere. Make sure you do not get them confused.

Aquifer A permeable rock that can store and transmit water.

Exam tip

Note you will have to be aware of all of these factors at a variety of scales: hill slope and drainage basin (see the following section) and global (see the previous section).

Evapotranspiration The combined water gain to the atmosphere by evaporation and transpiration.

Knowledge check 2

Identify and explain the factors that determine transpiration rates.

Condensation and cloud formation

Condensation is the process by which water vapour in the air is changed into liquid water. It occurs when saturated air is cooled, usually by a rise in altitude, below the dew point. Condensation is crucial to the water cycle because it is responsible for the formation of clouds. These clouds may produce precipitation, which is the primary route for water to return to the Earth's surface within the water cycle.

Precipitation

Precipitation is water released from clouds in the form of rain, freezing rain, sleet, snow or hail. Most precipitation falls as rain. For precipitation to happen the water droplets must condense on even tinier dust, salt or smoke particles, which act as a nucleus (condensation nuclei). Water droplets may grow as a result of additional condensation of water vapour when the droplets collide. If enough collisions occur to produce a droplet with a fall velocity that exceeds the cloud updraft speed, then it will fall out of the cloud as precipitation.

Another mechanism (known as the Bergeron-Findeisen process) for producing a precipitation-sized drop is through a process that leads to the rapid growth of ice crystals at the expense of the water vapour present in a cloud. These crystals may fall as snow, or melt and fall as rain.

Sublimation and de-sublimation

Sublimation refers to the conversion between the solid and gaseous phases of matter, with no intermediate liquid stage. It is most often used to describe the process of snow and ice changing into water vapour in the air without first melting into water. The opposite of sublimation is 'de-sublimation' (sometimes referred to as **deposition**), where water vapour changes directly into ice — such as hoar frost on trees.

Snow melt

Runoff from snowmelt is a major component of the global movement of water, although its importance varies greatly geographically. In warmer climates it does not directly play a part in water availability. In the colder climates, however, much of the springtime runoff and flow in rivers is attributable to melting snow and ice.

Drainage basins as open systems

Other aspects of the water cycle are best examined at a smaller scale: the drainage basin hydrological cycle. Inputs include energy from the sun and precipitation. Outputs include evaporation and transpiration (evapotranspiration), water percolating into deep groundwater stores and runoff into the sea. Stores can take place in a number of locations — on vegetation, on the ground, in the soil and in the underlying bedrock. Transfers take place between any of these stores and ultimately into the channels of the rivers of the drainage basin. Drainage basins are bounded by high land beyond which any precipitation will fall into the adjacent drainage basin. The imaginary line that separates adjacent drainage basins is called the **watershed**.

Dew point The temperature at which a body of air at a given atmospheric pressure becomes fully saturated. If an unsaturated body of air is cooled, a critical temperature will be reached where its **relative humidity** becomes 100% (i.e. saturated).

Exam tip

All of these processes represent transfers or flows within the water cycle system — be clear as to the direction of each of them, for example, evaporation is from land/sea to atmosphere. You may want to draw a diagram of them.

Drainage basin The catchment area from which a river system obtains its supplies of water.

Content Guidance

Drainage basin terminology

Groundwater store is water that collects underground in pore spaces in rock.

Groundwater flow is the movement of groundwater. This is the slowest transfer of water within the drainage basin and provides water for the river during drought.

Infiltration is the movement of water from the surface downwards into the soil.

Interception is the process by which precipitation is prevented from reaching the soil by leaves and branches of trees as well as by smaller plants such as grasses.

Overland flow (or surface runoff) is the movement of water over saturated or impermeable land.

Percolation is the downward movement of water from soil to the rock below or within rock.

Runoff is all the water that flows out of a drainage basin.

Stemflow is the water that runs down the stems and trunks of plants and trees to the ground.

Throughfall is the water that drips off leaves during a rainstorm.

Throughflow is the water that moves downslope through soil.

All of these flows lead water to the nearest river. The river then transfers water by channel flow. The amount of water that leaves the drainage basin is its **runoff**. The runoff of a river is measured by its **discharge**. For any river at a given location, this is calculated by the following:

Discharge (Q) = average velocity (V) × cross sectional area (A)

The unit is cumecs, measured in m^3s^{-1}.

The concept of the water balance

This refers to when inputs of precipitation (P) are balanced by outputs in the form of evapotranspiration (E) and runoff (Q) together with changes to the amounts of water held in storage within the soil and groundwater (ΔS):

$P = E + Q + \Delta S$

When precipitation exceeds evapotranspiration, this produces a water surplus. Water infiltrates into the soil and groundwater stores (passing below the water table). When pore spaces are saturated, excess water contributes to surface runoff. When evapotranspiration is greater than precipitation, evapotranspiration demands are met by water being drawn to the surface of the soil by capillary action. Groundwater stores are depleted and runoff tends to be reduced.

The flood (storm) hydrograph

A flood (storm) **hydrograph** is the graph of the discharge of a river leading up to and following a 'storm' or rainfall event — its runoff variation. A hydrograph is important because it can help predict how a river might respond to a rainstorm, which in turn

Exam tip

You should be able to consider the factors that affect each of these flows and stores. For example, infiltration is affected by the rate of precipitation, soil type, antecedent rainfall, vegetation cover and slope.

Knowledge check 3

Explain *two* ways in which overland flow is created.

Hydrograph A graph of river discharge against time.

can help in managing the river. When water is transferred to a river quickly the resultant hydrograph is described as **flashy**. This means that the river responds very quickly to the storm and often leads to flooding.

A number of natural variables within a drainage basin can have an effect on the shape of the storm hydrograph:

- **Antecedent rainfall:** rain falling on a ground surface that is already saturated will produce a steep rising limb and a shorter lag time.
- **Snowmelt:** large amounts of water are released, greatly and rapidly increasing discharge, especially if the ground surface is still frozen, as this reduces infiltration.
- **Vegetation:** in summer, deciduous trees have more leaves so interception is higher, discharge is lower and lag times are longer.
- **Basin shape:** water takes less time to reach the river in a circular drainage basin than in an elongated one.
- **Slope:** in steep-sided drainage basins water gets to the river more quickly than in an area of gentle slopes.
- **Geology:** permeable rocks allow percolation to occur, which slows down the rate of transfer of water to the river. Impermeable rocks allow less percolation and have greater amounts of overland flow, and hence greater discharges and shorter lag times.

Factors affecting change in the water cycle

Natural variations

Flash floods

These have a timeline measured in minutes or hours, but their effects can be devastating. On a hydrograph they have an extremely steep rising limb and a very short lag time. The recession limb is equally steep and the river returns to normal flow conditions in a matter of hours.

Single event floods

These result from a single input of additional water, such as a period of cyclonic rainfall, into the catchment. They have a timeline measured in hours and days from the onset of rainfall to the flood peak, and for the recovery back to normal conditions.

Multiple event floods

These result from a number of closely spaced single event floods in the same catchment such that they develop a very long timescale measured in weeks, or even months of continuous flooding. These are often the result of repeated frontal/cyclonic rainfall events, sometimes coinciding with snowmelt.

Seasonal floods

These occur where there is either seasonal rainfall or seasonal snowmelt on a massive scale. An example is the flooding associated with the Asian monsoon, in areas such as the Indus Valley in Pakistan and the floodplains of Bangladesh.

Flood A body of water that rises to overflow land that is not normally submerged.

Knowledge check 4

Explain how urban growth can affect a hydrograph.

Monsoon A seasonal reversal of wind direction that brings a period of intense rain to the area affected.

Knowledge check 5

Explain how variations in precipitation can impact the water cycle.

Human impacts

Land use change: deforestation

Deforestation removes water-absorbent forests, which trap and transpire rainfall, and replaces them with arable or grazing land. Consequently there will be a significant increase in both the volume of water reaching a river and the speed with which it travels. However, it has also been said that the impact of deforestation overall is overstated, with the main impact being limited to the tree felling period, when tracks are being driven through the forest and heavy machinery compacts the soil, causing additional overland flow. When the land reverts to scrub or pasture, runoff patterns return to their pre-deforestation state.

Farming practices: arable landscapes

It has been suggested that more emphasis on arable farming has created a greater flood risk. However, this impact varies according to the seasons. In late autumn, winter and early spring, crops are dormant and the soil is relatively bare. Rain falling on these surfaces is not intercepted by vegetation, and hence overland flow rates are relatively high. By contrast, in late spring and summer arable landscapes have growing and fully established crops that can intercept much greater proportions of the rainfall and thereby reduce peak flows, extending lag times.

Water abstraction

People all over the world make great use of the water in underground aquifers. In some places they pump water out of the aquifer faster than nature replenishes it. In these cases, the excessive pumping can lower the water table, below which the soil is saturated. Thus wells can 'go dry' and become useless. However, in places where the water table is close to the surface and where water can move through the aquifer at a high rate, aquifers can be replenished artificially.

Climate change

The likelihood of human-induced climate change may affect several of the factors given above. For example, in southern Asia the seasonal monsoon rains could be affected in both magnitude and timing, although the direction of those changes is uncertain. Some also suggest that tropical storms will be more frequent and more severe. In the UK, more intense rainstorms are likely to bring about an increased incidence of surface water flooding.

The carbon cycle

The distribution and size of carbon stores

Most of Earth's carbon, about 65,500 billion metric tonnes, is stored in rocks and the soil above it (the lithosphere). The remainder is in the hydrosphere (the oceans), atmosphere, the cryosphere (within the permafrost) and the biosphere (plants and animals). Carbon flows between each of these stores in a complex set of exchanges called the carbon cycle (Figure 1). Any change in the cycle that shifts carbon out of one store puts more carbon in the other stores. It is now widely accepted that changes which put more carbon gases into the atmosphere result in warmer temperatures on Earth and hence the carbon cycle has a close connection to climate change.

> **Exam tip**
>
> Shorter examination questions will often ask you to separate out physical (natural) and human factors, but longer synoptic questions will ask you to combine them.

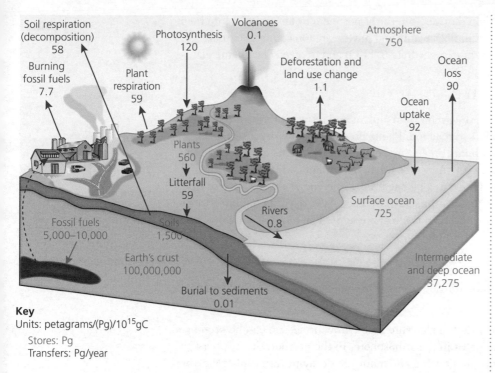

Key
Units: petagrams/(Pg)/10^{15}gC
Stores: Pg
Transfers: Pg/year

Figure 1 The carbon cycle

The carbon cycle The units of the carbon cycle are massive and confusing in much literature. This should help: 1 petagram of carbon per year (PgC/yr) = 1 gigatonne of carbon per year (1 GtC/yr) = 1 billion (10^9) tonnes = 10^{15} grams of carbon per year.

Factors driving change in the magnitude of carbon stores

Weathering

The movement of carbon from the atmosphere to the lithosphere begins with rain. Atmospheric carbon combines with water to form a weak carbonic acid that falls to the surface in rain. The acid dissolves rocks — a form of chemical weathering — and releases calcium, magnesium, potassium and sodium ions. Plants, through their growth, also break up surface granite, and microorganisms hasten the weathering with enzymes and organic acids in the soil coupled with the carbonic acid and carbon dioxide in the water.

Carbon sequestration in oceans and sediments

Weathered calcium and bicarbonates are then washed down to the sea by rivers and used by microscopic marine life to form shells. The ocean algae also draw down carbon dioxide from the air and when these microflora die, their remnants and the shells from other organisms sink to the ocean floor and are compressed to form sediments of limestone and chalk. Carbon locked up in limestone can be stored for millions, or even hundreds of millions of years. Carbon has also been stored in rocks where dead plant material built up faster than it could decay to become coal, oil and natural gas. Carbon is also stored in compressed clay deposits called shales.

Photosynthesis, respiration, decomposition and combustion

Plants (on land) and phytoplankton (microscopic organisms in the ocean) are key components of the carbon cycle. During **photosynthesis**, plants absorb carbon dioxide (CO_2) and sunlight to create glucose and other sugars for building plant structures.

Knowledge check 6

Explain the role of volcanoes in the carbon cycle.

Exam tip

Be aware of the relative sizes and rates of the various carbon cycle processes (Figure 1). It is unlikely that you will remember the numbers involved, but an understanding of relative scale is important.

Phytoplankton also take carbon dioxide from the atmosphere by absorbing it into their cells. Using energy from the sun, both plants and phytoplankton combine carbon dioxide and water (H_2O) to form carbohydrate (CH_2O) and oxygen. The chemical reaction looks like this:

$$CO_2 + H_2O + energy \rightarrow CH_2O + O_2$$

Four things can happen to move carbon from a plant and return it to the atmosphere, but all involve the same chemical reaction. Plants break down the sugar to get the energy they need to grow. Animals (including people) eat the plants or plankton, and break down the plant sugar to get energy (**respiration**). Plants and plankton die and decay (**decomposition**), and are eaten by bacteria, at the end of the growing season, or natural fire consumes plants (**combustion**). In each case, oxygen combines with carbohydrate to release water, carbon dioxide and energy. The basic chemical reaction looks like this (a reverse of the formula above):

$$CH_2O + O_2 \rightarrow CO_2 + H_2O + energy$$

In all four processes the carbon dioxide released in the reaction usually ends up in the atmosphere.

The carbon cycle is so tightly tied to plant life that a growing season can be seen by the way carbon dioxide fluctuates in the atmosphere. In the northern hemisphere winter, when few land plants are growing and many are decaying, atmospheric carbon dioxide concentrations climb. During the spring, when plants begin growing again, concentrations drop. It is as if the Earth is breathing.

Other forms of sequestration

Natural sequestration includes the following:
- **Peat bogs:** by creating new bogs, or enhancing existing ones, carbon can be sequestered.
- **Reforestation:** the planting of trees on marginal crop and pasture lands will incorporate carbon from atmospheric carbon dioxide into biomass.
- **Wetland restoration:** 14.5% of the world's soil carbon is found in wetlands, while only 6% of the world's land is composed of wetlands.

Human-induced sequestration includes the following:
- **Urea fertilisation:** fertilising the oceans with urea, a nitrogen-rich substance, which encourages phytoplankton growth.
- **Bio-energy with carbon capture and storage (BECCS):** where carbon is captured in power stations and stored underground.
- **Biochar:** the addition of charcoal to a soil.

Changes in the carbon cycle over time

Natural variations

Wild fires

What is a wild fire? Some define it as one caused by nature, such as a lightning strike. However, only 10% of wild fires are started this way. An alternative view is that a wild fire is one that has been started by humans, but gone out of control.

> **Exam tip**
> Make sure you can define each of the terms photosynthesis, respiration, decomposition and combustion.

> **Sequestration** The process of capture and long-term storage of atmospheric carbon dioxide by either natural or human-induced means.

> **Knowledge check 7**
> Explain how carbon exploitation (such as in power stations) and carbon capture can work together.

It appears that forest fires can release more carbon into the atmosphere than the forest can capture and that this may be a growing problem as the number of wild fires increases. Every year wild fires burn 3–4 million km^2 of the Earth's land area, and release tonnes of carbon into the atmosphere in the form of carbon dioxide. However, after a fire, new vegetation moves onto the burned land and over time reabsorbs much of the carbon dioxide that the fire had released.

Volcanic activity

Carbon is emitted to the atmosphere through volcanoes. Earth's land and ocean surfaces sit on several moving crustal plates. When the plates collide, one sinks beneath the other, and the rock it carries melts under the extreme heat and pressure. The heated rock recombines into silicate minerals, releasing carbon dioxide. When volcanoes erupt, they vent the gas to the atmosphere and cover the land with fresh silicate rock, to begin the cycle again. At present, volcanoes emit between 130 and 380 million metric tonnes of carbon dioxide per year.

Human impacts

Hydrocarbon fuel extraction and burning

Humans have interfered with the carbon cycle where fossil fuels have been mined from the Earth's crust and subsequently burned. The age of the organisms and their resulting fossil fuels is typically millions of years. Fossil fuels contain high percentages of carbon and include coal, oil and natural gas. They range from volatile materials with low carbon/hydrogen ratios, such as methane, to liquid petroleum and non-volatile materials composed of almost pure carbon, such as anthracite coal.

The use of fossil fuels raises serious environmental concerns. Their burning produces around 21 Pg of carbon dioxide per year, but it is estimated that natural processes can only absorb about half of that amount, so there is a net increase of about 8 Pg of atmospheric carbon dioxide per year. Carbon dioxide is one of the greenhouse gases that enhances atmospheric heating and contributes to climate change.

Land use changes

The Intergovernmental Panel on Climate Change (IPCC) estimates that land use change, such as the conversion of forest into agricultural land, contributes a net 1.6 Gt carbon per year to the atmosphere. Various types of land-use change can result in changes to carbon stores:

- conversion of natural ecosystems to permanent croplands
- conversion of natural ecosystems for shifting cultivation
- conversion of natural ecosystems to pasture
- abandonment of croplands and pastures
- deforestation — the harvesting of timber
- wetland clearance
- establishment of tree plantations (afforestation)

Deforestation: when forests are cleared for conversion to agriculture or pasture, a very large proportion of the above-ground biomass may be burned, releasing most of its carbon rapidly into the atmosphere. Forest clearing also accelerates the decay of dead wood and litter, as well as below-ground organic carbon. Local climate and soil

> **Exam tip**
>
> The impact of increased concentrations of carbon dioxide on world climate is well documented. You should be aware of what the IPCC says on this topic.

conditions will determine the rates of decay; in tropical moist regions, most of the remaining biomass decomposes in less than 10 years. Some carbon or charcoal also accretes to the soil carbon pool.

Wetland clearance: when wetlands are drained for conversion to agriculture or pasture, soils become exposed to oxygen. Carbon stocks, which are resistant to decay under the anaerobic conditions prevalent in wetland soils, can then be lost by aerobic respiration.

Farming practices

Examples of impact include the following:

- Cropland soils can lose carbon as a consequence of soil disturbance such as tillage. Tillage increases aeration and soil temperatures, making soil aggregates more susceptible to breakdown and organic material more available for decomposition.
- Soil carbon content can be protected and even increased through alteration of tillage practices, crop rotation, residue management, reduction of soil erosion, improvement of irrigation and nutrient management.
- Heavy livestock grazing alters the ground cover and can lead to soil compaction and erosion, as well as alteration of nutrient cycles and runoff. Avoiding overgrazing can reduce these effects.
- Rice cultivation and livestock have been estimated to be the two primary sources of methane. Alteration of rice cultivation practices, livestock feed and fertiliser use are therefore potential management practices that could reduce methane sources.

The carbon budget

Carbon dioxide is the single most important anthropogenic greenhouse gas in the atmosphere, contributing approximately 65% to **radiative forcing** by greenhouse gases. It is responsible for the majority of the increase in radiative forcing over the past decade. The current **carbon budget** shows a net gain of 4.3 Gt of carbon per year in the atmosphere. The rising levels of carbon dioxide and other greenhouse gases in post-industrial times are fuelling fears of climate change through atmospheric warming.

Atmospheric carbon dioxide reached 142% of the pre-industrial level in 2013, primarily because of emissions from combustion of fossil fuels and cement production. Relatively small contributions to increased carbon dioxide come from deforestation and other land-use change, although the net effect of terrestrial biosphere fluxes is as a sink, at 2 Gt per year. The average increase in atmospheric carbon dioxide from 2003 to 2013 corresponds to 45% of the carbon dioxide emitted through human activity, with the oceans and the terrestrial biosphere removing the remaining 55%.

Water, carbon, climate and life on Earth

The relationship between the water cycle and the carbon cycle

Both the carbon and water stores and cycles play a key role in supporting life on Earth, largely through their influence on climate. There are clear relationships between the water and carbon cycles in the atmosphere. Furthermore, the various

Radiative forcing The difference between incoming solar radiation (insolation) absorbed by the Earth and the energy radiated back out into space.

Carbon budget The balance of exchanges between the four major stores of carbon.

feedback mechanisms within and between the cycles are strongly linked to climate change. Most scientists now agree that climate change will significantly impact life on Earth.

In simple terms, the relationships between the three components of the water cycle, carbon cycle and climate change are as follows:

- Changes in the carbon cycle are the main factors causing climate change.
- Climate change is having, and will continue to have, an effect on the water cycle, such as more evaporation and/or more precipitation in certain parts of the world.
- Climate change is having, and will continue to have, an effect on the carbon cycle, such as the release of more carbon dioxide from the permafrost areas of the world as they warm.

Humans have to either **adapt** to the water cycle-related outcomes of climate change, such as increased rates of ice cap melting, flooding and drought, or **mitigate** these impacts by managing the carbon cycle, or both.

Adaptation Changing lifestyles to cope with climate change.

Mitigation Reduction in the output/amount of greenhouse gases.

The nature of climate change

Climate change can be assessed across a variety of time scales:

- **Long-term (or geological):** over several hundreds of thousands to millions of years
- **Medium-term (or historical):** within the last few thousand years
- **Short-term (or recent):** within the last few decades

Long-term climate change

The best evidence for this comes from ice cores in Greenland and Antarctica. An ice core taken from the 3,200 m deep East Antarctic ice sheet has recorded the climate of the last 800,000 years. Air bubbles trapped in the ice contain atmospheric carbon dioxide and the ice itself preserves a record of oxygen isotopes. Low concentrations (180 parts per million (ppm)) of carbon dioxide occur naturally during glacial (cold) periods and high concentrations (280 ppm) during interglacial (warm) periods. It is clear that atmospheric carbon dioxide levels are higher now (390 ppm) than at any time for over half a million years.

Medium-term climate change

Studies related to changes in vegetation provide strong evidence of climate change over this time period. Pollen extracted from sediment cores in peat bogs and lake beds records the ecology of the past. Pollen grains are preserved in waterlogged sediments that are anaerobic, or oxygen-free. Each plant species has a distinctively shaped pollen grain, which can be easily identified. In the UK, pollen sequences have shown that ecosystems have changed in response to climate change. Tundra ecosystems were present in past glacial periods, whereas forest gradually colonised areas as interglacial conditions developed.

Recent climate change

In January 2014 the IPCC published its fifth *Assessment Report on Climate Change* (*AR5*). The report draws on the work of over 800 scientists and cites more than 9,000 scientific publications. It states that it is 'virtually certain' humans are to blame for 'unequivocal' global warming.

The report also states:

> Warming of the climate system is unequivocal, and since the 1950s, many of the observed changes are unprecedented over decades to millennia. The atmosphere and ocean have warmed, the amounts of snow and ice have diminished, sea level has risen, and the concentrations of greenhouse gases (GHGs) have increased.

Other key findings of *AR5* included:

- Atmospheric concentrations of the GHGs carbon dioxide, methane and nitrous oxide are at levels 'unprecedented in at least the last 800,000 years'.
- Since 1750, fossil-fuel burning has increased the carbon in the Earth's atmosphere by about 40%.
- More than half of all global warming since 1951 is attributed to human activity.

The causes of climate change

There is no single cause of climate change. On the very long timescales of glacial to interglacial cycles, the most common explanation is astronomical forcing. On medium timescales of hundreds to thousands of years, variations in the sun's solar output may fit the observed trends. The recent warming that the Earth has experienced in the last few decades is increasingly seen as being driven by human pollution of the atmosphere (known as anthropogenic warming, or the enhanced greenhouse effect). The IPCC's *AR5* report states:

> Total radiative forcing is positive, and has led to an uptake of energy by the climate system. The largest contribution to total radiative forcing is caused by the increase in the atmospheric concentration of CO_2 since 1750. Human influence on the climate system is clear. This is evident from the increasing GHG concentrations in the atmosphere, positive radiative forcing, observed warming, and understanding of the climate system.

The impacts of climate change

Although it is not directly stated by the AQA specification, you should be aware of the impacts of climate change in a range of contexts, and especially those that may concern the water cycle and affect life on Earth. Some examples you could study are:

- the impact of climate change on the Arctic and/or Antarctica
- the impact of climate change on the permafrost, especially in northern latitudes (see also page 53, Glacial systems and landscapes)
- the impact of climate change on low-lying islands in the Pacific Ocean

Human interventions in the carbon cycle

There is general agreement that climate change needs to be addressed, but much less agreement on how this can be achieved. In general terms, there are two approaches: **mitigation** and **adaptation**.

Exam tip

You could research *AR5* in more detail. There is plenty of information online.

Astronomical forcing The changing of the surface temperature of the Earth over time due to variations in the orbit and axis tilt of the Earth.

Greenhouse effect The natural processes whereby outgoing thermal radiation is trapped by atmospheric gases such as carbon dioxide and methane.

Enhanced greenhouse effect The increased impact of greater amounts of greenhouse gases caused by human activity.

Mitigation refers to the reduction in the output of GHGs and/or increasing the size and amount of GHG storage or sink sites. Examples of mitigation are:

- setting targets to reduce greenhouse gas emissions
- switching to renewable sources of energy
- 'capturing' carbon emissions and/or storing or burying them (sequestration)

Adaptation refers to changing our lifestyles to cope with a new environment rather than trying to stop climate change. Examples of adaptation are:

- developing drought-resistant crops
- managing coastline retreat in areas vulnerable to sea-level rise
- investing in better quality freshwater provision to cope with higher levels of drought

You should be aware of the variety of interventions in the carbon cycle at a variety of scales: global, regional, national and local.

Global interventions include agreements such as the Kyoto Protocol. This set specific, legally binding targets for pollution mitigation and proposed schemes to enable governments to reach these targets. Most governments agreed that by 2010 they should have reduced their atmospheric pollution levels (of GHG emissions) to those prior to 1990. The Kyoto Protocol came into force in February 2005 and by 2006 was ratified by 183 countries. However, this agreement expired in 2012. There have been several attempts to update Kyoto (the Gleneagles Action Plan (2005), the Bali Road Map (2007), and the Copenhagen Accord (2009)), but none has been fully successful. A further conference, the United Nations Climate Change Conference, or COP21, took place in Paris in late 2015.

Outcomes of the COP21 meeting in Paris, 2015

The COP21 agreement contains a number of specific elements that together comprise the Paris commitment:

- A temperature increase of 2°C is to be avoided, and efforts made to limit the increase to a lower target of 1.5°C.
- GHG emissions will be allowed to rise for now, but with sequestration aimed for later this century in order to keep within scientifically determined GHG budgets.
- Emissions targets will be set by countries separately, but reviewed every 5 years, and after each review emissions levels decreased meaningfully.
- Accurate emissions records will be kept and made available to all other countries.
- Wealthy countries will share science and technology relating to low GHG emissions routes to economic and social development.
- Wealthy countries will make affordable finance available for those poor nations most affected by anthropogenic climate change.
- Countries that have historically emitted a lot of GHGs (like the UK) will recognise the 'loss and damage' inflicted on poor countries because of climate change.

A **regional** approach is that of the European Union (EU) with its 20/20/20 vision, which states that there will be a 20% reduction in GHG emission, a commitment to 20% of energy coming from renewable sources and a 20% increase in energy efficiency, all by 2020. The EU has also offered to increase its emissions reduction to 30% by 2020 if other major emitting countries in the developed and developing worlds commit to undertake their fair share of a global emissions reduction effort.

Content Guidance

At a **national** scale, the UK government introduced the Climate Change Act of 2008. This Act set a legally binding target for the UK to reduce GHG emissions by 80% compared with 1990 levels by 2050, with an interim target of 26% by 2020 (which has subsequently been increased to 34%). It also established a series of national carbon budgets and created the independent Committee on Climate Change to advise the government and report on progress.

At a **local** scale, individuals can respond to climate change by improving home insulation, recycling, using energy more wisely (for example, using smart meters), using public transport or car sharing schemes, and calculating personal carbon footprints.

Knowledge check 9

Carbon trading systems are becoming more common around the world. Describe and explain one carbon trading system.

Case studies

You are required to have studied two case studies:
1 A case study of a river catchment(s) at a local scale to illustrate and analyse the key themes of the water cycle system and consider the impact of precipitation upon drainage basin stores and transfers. There is a strong recommendation that you carry out fieldwork, or use field data, in such an environment. You should also consider how the changes within the drainage basin impact flooding and flood control and/or sustainable water supply. With these in mind, good examples would include upland catchments within the British Isles.
2 A case study of a tropical rainforest environment to illustrate and analyse the key themes of both the water and carbon cycle systems. You should also research the relationship of these two cycles to environmental change and human activity. With these in mind, good examples would include rainforests within the Amazon region of South America, southeast Asia and central Africa, where human activity (largely deforestation and new farming practices) is contributing to change in a significant manner.

Exam tip

The exam questions on these case studies are likely to use one or more of the following words: sustainable, resilience, mitigation and adaptation. Make sure you understand these terms.

Summary

After studying this topic, you should be able to:
- understand the principles of systems in physical geography and know the meaning of terms such as inputs, outputs, flows and transfers, stores and components, dynamic equilibrium and feedback mechanisms
- describe the global distribution of stores within the water cycle, and understand the factors driving change in the magnitude of these stores
- understand how drainage basins act as open systems, and the processes that cause change within them
- explain how both natural and human factors can affect both drainage basins and the water cycle as a whole

- describe the global distribution of stores within the carbon cycle, and understand the factors driving change in the magnitude of these stores
- explain how both natural and human factors can affect the carbon cycle at a variety of scales
- discuss the key roles of the carbon and water cycles in supporting life on Earth, with particular reference to climate
- evaluate the relationships between the water and carbon cycles, and how each of them contributes to climate change and has implications for life on Earth
- evaluate human interventions in the carbon cycle that are designed to influence carbon transfers and mitigate the impacts of climate change

■ Hot desert systems and landscapes

Deserts as natural systems

The deserts of the world are areas in which there is a substantial deficit of water, predominantly because they receive small amounts of precipitation. They have high levels of aridity; some areas of the world receive 100 mm, or even less, precipitation in any one year. It is this shortage of moisture, often exacerbated by high temperatures, that determines many of the characteristics of the soils, vegetation, the animals, the landscapes and human activities of such areas. Deserts can also be defined by the water balance — the difference between water from precipitation and losses due to evapotranspiration and changes in groundwater storage. Arid regions have an overall deficit over a year.

Desert environments are natural, open systems with inputs and outputs and flows between the two.

Inputs consist of:
■ energy provided by the sun through insolation, wind and flows of water
■ sediment provided from weathering, mass movement and erosion by both wind and water
■ changes resulting from human activity, such as desertification

Outputs consist of:
■ desert landforms both from erosion and deposition
■ accumulations of sand over thousands of years
■ losses of water through evapotranspiration and river flows
■ sand particles blown away by winds

Stores are essentially the extensive areas of sand (ergs) that occupy about 30% of deserts, and residual areas of water (rivers and lakes), which are small in number. **Transfers** result from the actions of wind and water.

You should refer back to the section on **feedback** and **dynamic equilibrium** in the earlier part of this guide (see page 6).

The global distribution of hot desert environments

One-third of the Earth's land surface is arid in some form (including semi-arid) — and home to 20% of the world's population. These areas are often referred to as 'deserts', of which the Sahara, Namib, Kalahari, Atacama, Patagonian, Arabian, Thar, Mojave, Sonora and Australian deserts are the best known.

Deserts are classified in a number of ways. Some refer to warm (or hot, or mid- and low-latitude) deserts, whereas there are also deserts where, due to either high latitude or high altitude, there are winter frosts (cold deserts). Likewise, coastal deserts, such as the Atacama and Namib, have very different temperatures and levels of humidity from those deserts of continental interiors. Equally, some deserts, such as those in

Landscape An expanse of land/scenery that can be seen in a single view. It covers all aspects of the view — both natural landforms and human-created features.

Insolation The heat energy of the sun (**in**coming **sol**ar radi**ation**).

Knowledge check 10

Describe the various types of precipitation in desert areas.
....

Arabia and Australia, have much less relief (mountains and valleys) than others. The mountain and basin deserts of southwest USA and Iran have much steeper relief and consequently have very different landscape features. Hot deserts can be classified as being extremely (hyper) arid (< 50 mm precipitation per year), arid (< 250 mm precipitation per year) and semi-arid (250 mm to 500 mm precipitation per year).

Exam tip

When asked to describe distributions or patterns of features, give an overview of common locational features rather than a list of locations.

Climate of hot desert environments

Air temperatures in hot deserts are characterised by their extremes. These occur both:

- **diurnally (between day and night):** the clear skies allow both intense insolation during the day (30°C+) and rapid heat loss to space at night (in some areas below 0°C). Diurnal ranges of over 25°C are common.
- **annually (throughout the year):** latitude is the most important factor in the annual range; the further away from the tropics, the greater the range

All deserts have a negative water balance, whereby evapotranspiration is in excess of precipitation. Several climatologists have attempted to devise a quantitative index, expressing the relationship between precipitation and evapotranspiration, that determines aridity. In those areas of the world where there is little precipitation annually or where there is seasonal drought, the calculation of potential evapotranspiration is used. The best-known aridity index was put forward by C. W. Thornthwaite:

$$\text{aridity index} = \frac{(100 \times \text{water surplus}) - (60 \times \text{water deficit})}{\text{potential evapotranspiration}}$$

A value of −20 to −40 indicates a semi-arid area, and a value less than −40 indicates an arid area.

Potential evapotranspiration The amount of water that could be evaporated or transpired from an area, given sufficient water for this to happen.

Desert vegetation and soils

Desert plants can be categorised as ephemeral, xerophytic, phreatophytic and halophytic:

- **Ephemeral** plants have a short life cycle and may form a fairly dense stand of vegetation immediately after rain. They evade drought, and when rain falls they develop vigorously and produce large numbers of flowers and fruit. The seeds then lie dormant until the next wet spell, when the desert blooms again.
- **Xerophytic** plants possess drought-resistant adaptations. Transpiration is reduced by means of dense hairs covering waxy leaf surfaces. They close their stomata to reduce water loss and either roll up or shed leaves at the beginning of the dry period. Some xerophytes — the succulents — store water in their structures. Most xerophytes have extensive shallow root networks to search for water.
- **Phreatophytic** plants have long tap roots, which penetrate downwards until they reach deep sources of groundwater.
- **Halophytes** are salt-tolerant plants, common in the vast salt plains in inland desert basins.

Exam tip

Research an example of each of these four types of vegetation.

Soils in hot deserts (**aridosols**) are coarse-textured, shallow, rocky or gravelly, with little organic matter. This is caused by the low plant productivity, which restricts the soil-building properties of microorganisms. Sometimes the accumulation of salts up through a soil progresses so far that hard surface or subsurface crusts develop, called **duricrusts**.

Causes of aridity

Hot deserts occur in four types of location where there is very little rain:

- latitudes dominated by dry, subsiding air
- inland, far from sources of moist, maritime air
- on coasts flanked by cold ocean currents or cold upwelling ocean water
- in the rain shadow of high mountain ranges

Influence of latitude

Many of the world's largest deserts (e.g. Sahara and Australian) occur between the latitudes of 15° and 35°N and S. These are areas where the air above them is dry and subsiding due to the Hadley cells, and the atmospheric pressure is high for much of the year. The surface winds in such deserts are therefore generally directed outwards (trade winds), towards areas of lower atmospheric pressure, and so little moisture is brought in by surface winds. As the air over the deserts subsides it is compressed and becomes warmer, so it is able to absorb additional water vapour, reducing further the potential for rain.

Inland deserts

These are far enough inland to be away from the influence of moist maritime air masses. Rainfall decreases rapidly away from the coast in all parts of the world except those close to the equator. The drying influence of a large land mass is referred to as continentality, and this applies to all big deserts, including the great tropical deserts of Arabia, Australia and the Sahara. In these hot, tropical deserts, the dryness caused by latitude is accentuated by continentality. The deserts of central Asia, including the Taklamakan and Gobi deserts of China and Mongolia, are in the interior of mid-latitude continental regions far from the oceans.

Ocean currents

The presence close offshore of cold upwelling water or a cold ocean current can cause coastal aridity in tropical and even in equatorial latitudes, such as the arid Horn of Africa, flanked by the cold Somali current. In fact, the western borders of all the great tropical or trade-wind deserts in both hemispheres are washed by cool ocean currents associated with the oceanic circulation cells, which flow clockwise in the northern and anticlockwise in the southern hemisphere. These include the Atacama and Kalahari deserts.

Rain shadow effect

Wherever ranges of hills or mountains lie close to the coast, forming a physical barrier to onshore winds, the incoming moist maritime air will be forced upwards. Moist air becomes cooler as it rises and expands, attains vapour saturation and sheds its condensed water vapour as rain or snow. The air then passes over the coastal ranges and flows downhill, becoming warmer and drier. The area inland of the coastal ranges, such as Patagonia in South America, is described as being in the rain shadow of the ranges — the air has already shed its moisture before passing over this land, which therefore gets little rain.

Knowledge check 11

Define the term Hadley cell and explain how they are formed.

Ocean currents Large-scale movements of water within the oceans that are part of the process of the transfer of heat from the equator to the North and South Poles.

Knowledge check 12

Winds known locally as the Chinook are often associated with the rain shadow effect. Describe and explain these winds.

Systems and processes

Weathering and mass movement

Mechanical weathering, involving the disintegration of rocks without any chemical change, takes place in hot deserts due to two main factors:

- the high rates of insolation
- the action of salt

Insolation weathering is the rupturing of rocks and minerals primarily as a result of large daily temperature changes that lead to temperature gradients within the rock mass. These are manifested in a number of ways:

- **Exfoliation (onion skin weathering)** is most noticeable on rocks with few joints or bedding planes (e.g. granite, massive sandstone). It occurs because the surface heat does not penetrate very deeply into these rocks. The outer layers expand and contract daily, but the interior does not. This differential leads to weaknesses parallel to the surface. Layers peel off the rock in an onion skin-like way.
- **Granular disintegration** occurs in rocks that comprise minerals of different colours. Darker minerals absorb more heat than lighter ones (e.g. black mica expands more than grey quartz) and so the rock breaks up grain by grain to produce sand-sized material.
- Those rocks that are microcrystalline (e.g. basalt) are subject to shattering (**thermal fracturing**) by constant diurnal expansion and contraction due to heating and cooling respectively.
- **Block separation** occurs predominently on well-jointed and bedded limestones, where the rock breaks up into blocks along these weaknesses.
- **Frost shattering** can occur in deserts where there is some free water and the nighttime temperatures fall below zero. The water gets into fractures in the rock, freezes at night and so expands. Continual freezing and thawing will result in fragments breaking off.

Many desert geographers now believe that these mechanical forms of weathering are encouraged by the introduction of water, and particularly so when water moves up through groundwater, bringing with it weathered salts from below ground. These may enhance mechanical forms of weathering but there may also be some chemical changes. Salt weathering operates in two ways:

1 When a solution containing salts is either cooled or evaporated, salt crystals will form and pressures accompanying this crystallisation can be great enough to exceed the tensile strength of the rocks in which the solution was contained.

2 Salt minerals expand when water is added to them. This change of state is called hydration. In the case of sodium sulfate and sodium carbonate the expansion may be as great as 300%. If these salts are in the rock, the pressures generated can break the rock.

With the general lack of water in the desert environment, most forms of mass movement rely on the natural force of gravity. Rock falls and occasional rockslides are the main form of movement whereby weathered or fragmented rock falls from high up a slope down on to the ground below.

Exam tip

When describing different types of weathering, try to link them to specific rock types, as not all rocks are weathered in the same way.

Knowledge check 13

Distinguish between mechanical weathering and chemical weathering.

Exam tip

Questions on this area of study will require an understanding of the distinctive nature of the processes in this environment. Links to heat and the lack of water are crucial here.

The role of the wind (Aeolian processes)

As deserts are so dry and vegetation cover so limited, there is little to protect the desert surface against the action of the wind. Wind is able to erode desert surfaces. It does this in two ways:

1 Deflation: loose, fine material is picked up by the wind and is transported and deposited elsewhere. It can create depressions in the desert floor (**deflation hollows**), e.g. the Qattara depression in Egypt, or it can remove fine sand from a surface, leaving behind coarser stones that blanket the surface to form a **reg** or **desert pavement**.

2 **Abrasion:** a sand-blasting effect, where fine material carried in the wind is blown at rocks. **Ventifacts** are rocks lying on the floor of deserts that have been shaped by the wind-driven sand. They usually have sharp edges and smooth sides. Other common landforms include **zeugen** and **yardangs**. The coarsest sand grain material, which has the greatest effect, cannot be lifted more than 1.5 m off the ground. This results in undercutting and fluting. **Zeugen** form where there are large areas of horizontal rock layers with vertical jointing. Abrasion erodes the weak joints and then softer layers are eroded underneath. Thus long ridges develop with a protective cap rock. **Yardangs** occur when the layers of rock lie at a steep angle to the surface and parallel to the prevailing wind. The less resistant rock layers are eroded more rapidly than the resistant ones, producing long rock ridges. These can be several kilometres long and hundreds of metres high.

Wind is able to **transport** material in three ways:

1 **Saltation:** medium-sized grains of sand (0.15–0.25 mm diameter) move in a series of hops across the surface. The wind picks up a grain to a height of a few centimetres, carries it a short distance and then drops it. Where the grain falls it can dislodge other grains that are then picked up, and the process continues.

2 **Suspension:** High-velocity winds can lift and carry fine silt and clay (< 0.15 mm diameter) high into the atmosphere. This can often be taken away from the desert area entirely. Saharan dust has reached places as far away as the UK.

3 **Surface creep:** coarser grains of sand (> 0.25 mm diameter) are rolled across the surface.

Landforms resulting from deposition

Deposition occurs when the wind can no longer move the desert material. These deposits can then be shaped by the wind. Only about one-third of the world's deserts are covered by wind-blown sand. However, great ergs (seas of sand) can form some of the most striking landscapes seen on Earth.

Sand dunes are formed by the wind's re-working of large deposits of sand. Loose sand is blown up the windward side of a dune. The sand particles then fall to rest on the downwind side (usually at an angle of about 34°), while more are blown up from the windward side. In this way a dune moves gradually downwind. The geometric

Deflation The picking up and blowing away of fine, loose material by the wind.

Exam tip

A good way to demonstrate that you know what each of these landforms looks like is to draw a sketch of it. Give it a go!

forms of dunes are varied and depend on the supply of sand, the nature of the wind regime, the extent of vegetation cover and the shape of the ground surface.

The typical crescent-shaped dune, with two horns pointing downwind, is called a **barchan**. It may be up to 30 m high and 400 m across, and may move downstream at a speed of 15 m a year. It is formed by there being a dominant prevailing wind direction with it being orientated with its axis at right angles to the wind direction. The horns extend downwind as there is less sand at the edges.

Seif dunes are longitudinal and lie parallel to the prevailing wind direction. They can be very long (> 100 km) and can reach 30 m in height. They are commonly 200–500 m apart. Some suggest that they are formed in areas where there is a seasonal change in wind direction that connects crescent (barchans) dunes. Others state that they can be attributed to some regularity in patterns of turbulence in the wind.

The role of water

Present and past river action are also important in the moulding of desert landscapes. Although rainfall quantities are low overall, substantial amounts of rainfall may occur from time to time. Moreover, many desert surfaces have a number of characteristics that enable them to generate considerable runoff from quite low rainfall intensities:

- Limited vegetation cover provides little organic matter on the surface to absorb water.
- The sparseness of the vegetation means that humus levels in the soil are low, and combined with minimal disturbance by plant roots, this makes the soil dense and compact.
- As there is virtually no plant cover to intercept rainfall, rain is able to beat down on the soil surface with maximum force.
- Fine particles, unbound by vegetation, are redistributed by splash to lodge in pore spaces and create a surface of much reduced permeability.

Consequently, as a result of all these factors, infiltration rates are very low and overland flow is highly likely. **Channel flash floods** and sheetfloods are common in deserts following short periods of intense rainfall that cannot infiltrate. They can be devastating in their impact despite being short-lived.

There is a variety of water sources in desert areas:

- **Exogeneous** rivers are large perennial rivers that originate outside the desert, for example the River Nile's main tributary, the Blue Nile, which originates in the Ethiopian Highlands.
- **Endoreic** rivers are also perennial and flow into an inland drainage basin in the desert. For example, the River Jordan flows into the Dead Sea, which, at 422 m below sea level, has no outlet.
- **Ephemeral** rivers are created by heavy rainstorms in a desert. The rain cannot infiltrate into the baked desert surface and flows over land as sheet flow or in what are normally dry valleys (wadis). After the storm, the water quickly evaporates or soaks into the wadi floor.

The action of water has produced some distinctive landforms.

Wadis are dry stream channels or valleys found in deserts. They are formed either by ephemeral or endoreic rivers. When the rivers are flowing, they can carry huge

Exam tip

A good way to demonstrate that you know what each of these landforms looks like is to draw a sketch of it. Give it a go!

Exam tip

You will be asked to describe landforms. Refer to size, shape, nature of sediments and field relationship (i.e. where the landform lies in relation to the landscape).

Knowledge check 14

Another type of dune is a star dune. Describe its characteristics and explain its formation.

Sheetfloods These floods remove thin layers of surface material evenly from an extensive area of gently sloping land, by broad continuous sheets of running water rather than by streams flowing in channels.

amounts of sediment. They are able to erode vertically to form steep sides. When the rivers dry up this sediment is deposited on the floor of the wadi, to give it its typical flat floor. In some parts of the Sahara there are extensive systems of wadis that could not have been created by present-day water erosion. It is likely that they were excavated by water erosion at a time when the rainfall was heavier than at present. Some believe they were formed during the ice ages — the periods of advancing ice sheets in Europe coincided with periods of heavier rainfall (pluvials) in North Africa.

Alluvial fans (also known as **bahadas** or **bajadas**) are cones of sediment that occur between mountain fronts and low-lying plains. Their size is variable — small ones may have a radius of only a few tens of metres while larger ones may be more than 20 km across, and 300 m thick at the apex. The larger ones consist of a gently sloping plain stretching from the mountain edge into the desert. They form where rivers emerge from a confined upper mountain valley, or wadi (or canyon). At such a point the river can spread out, decreasing its velocity so that deposition occurs from a sediment-rich flow.

Other aspects of arid landscape development

Desert slopes

Slope profiles in desert areas are often visually dramatic. Their form is often analysed in terms of an idealised profile: namely, a cliff or mountain front, a straight segment and a **pediment**. Pediments may coalesce and cover extensive areas and from them may rise residual outliers called **mesas** and **buttes**.

Pediments are extensive, low-angled (1–7°) rock surfaces at the base of desert mountain fronts or cliffs. They may have a thin veneer of rock debris. Their origin is uncertain, but it is thought that a combination of the retreat of the mountain front by weathering and a removal of debris by sheets of water or sheetfloods formed them.

Rising above the pediments are often steep-sided rocky hills called **inselbergs**, formed as a result of gradual slope retreat produced by weathering and erosion at the break of the slope. They can be up to 600 m high and form spectacular features in many deserts. Inselbergs occur in many parts of the world, including Africa and Australia; Uluru (Ayers Rock) is perhaps the most famous such rock in the world.

Mesas and **buttes** are particular forms of inselbergs, associated with the USA. They are formed in areas of sedimentary rock with horizontal bedding planes. Water has eroded the land around them, leaving some resistant portions, with a resistant cap rock standing out from the surface. Mesas are steep-sided and flat topped. Buttes are similar, but are much more eroded and are pillar-like in appearance. Their lower slopes are often covered in scree — the result of weathering.

Away from the steep relief of these mountains, sometimes shallow salt lakes, called **playas**, develop in lowland basins. Sometimes the water is evaporated from these lakes to leave extensive salt pans. For example, the extensive salt flats known as Lake Bonneville can be found in the vicinity of the Great Salt Lake in Utah, USA.

Knowledge check 15

Badlands are features of semi-arid landscapes in the USA. Describe their characteristics and explain their formation.

Exam tip

Again, you will be asked to describe landforms. Refer to size, shape, nature of sediments and field relationship (i.e. where the landform lies in relation to the landscape).

Knowledge check 16

Make three lists: desert landforms created by wind; desert landforms created by water; and distinctive desert slope landforms.

Desertification

Desertification is the degradation of land in arid, semi-arid, and dry sub-humid areas. It is caused primarily by human activities and climatic variations. Desertification does not refer to the expansion of existing deserts. It occurs because dryland ecosystems, which cover over one third of the world's land area, are extremely vulnerable to over-exploitation and inappropriate land use.

(UNFAO)

The United Nations Convention to Combat Desertification (UNCCD) also defines desertification as 'land degradation in arid, semi-arid and dry sub-humid areas resulting from various factors, including climatic variations and human activities.' In turn, land degradation is defined as 'the reduction or loss of the biological or economic productivity of drylands'.

The main **physical cause** is climate change. In the vulnerable areas, the following changes are taking place:

- There is less rainfall and, what there is, is less reliable. This means farmers find it difficult to plan ahead. There is an increase in the frequency and intensity of droughts.
- There are higher temperatures. This increases evaporation, reduces condensation and leads to less rainfall.
- As there is less rainfall, rivers dry up and the water table falls.

The prospects of even less rainfall in the future have featured in all the IPCC *Assessment Reports* (*ARs*) produced so far.

There are several **human causes:**

- The main human cause is population growth. It is estimated that over 1 billion people live in areas at risk. The numbers of these people are increasing because of high birth rates and an influx of refugees from conflict or drought-hit areas. This population growth has led to increasing livestock numbers, which in turn leads to overgrazing.
- Farming of marginal or low-quality land is causing desertification worldwide. Farmers are clearing marginal land and overusing it, which takes away the richness in the soil — they are not letting it have a fallow period in order to replenish itself before farming resumes.
- Deforestation is causing desertification to occur. People are cutting down trees to use them as a source of fuel for the growing populations. Once all these trees are cut down there is nothing to protect the soil. Therefore, it turns to dust and is blown away by the wind.
- Unsuitable irrigation systems are commonly used in poorer areas. Farmers often use canal irrigation and other poor techniques because of the lack of water. This type of irrigation causes a build-up of salt in the soil, which makes it difficult for plants to grow due to the salinity.

Exam tip

You could research desertification in the latest IPCC report, *AR5*. See what it says about the prospects of the case study area you examine in detail.

- The changing nature of world trade in food has led to an increase in the growing of cash crops for markets in the more developed world (such as coffee in Ethiopia). This means that the best land is used for these crops and unsuitable marginal land is used for growing food crops.
- Several areas suffering from the other causes of desertification are often caught up in political instability and conflict. These cause poor land management practices to come to the fore, or abandonment.

All of these factors combine and lead to further reductions in vegetation on the land. Soil is then exposed to the wind and rain, which in turn leads to soil erosion because the root mat that holds the soil together is gone. This loss of soil then leads to desertification — an example of how physical and human feedback processes operate in a cumulative, negative and damaging way.

Managing the world's dryland environments is perhaps one of the most challenging and pressing development problems of today. Long-term strategies for eradicating poverty in the world's drylands will only succeed when the natural resources on which the people who live there depend are used sustainably. Programmes for protecting the dryland environment will accomplish their aims only when they also address the day-to-day pressures of poverty.

In general terms, there are two approaches to managing the drylands:

1 To mitigate against the drought. Drought cannot be prevented, although scientists have experimented with cloud seeding using silver iodide pellets to bring on rainstorms. Dealing with drought mainly involves setting up water storage schemes and increasing community preparedness.

2 To adapt to the occurrence of the drought and modify **vulnerability**, thereby increasing **resilience**. Scientists use satellite imagery to measure the progress of the rains in order to predict drought areas before people begin to suffer due to crop failure. This provides some accuracy to warnings, but often aid and government agencies lack the resources to respond in time. Longer-term drought aid in the form of irrigation schemes and education for farmers on water conservation techniques is perhaps more useful. Another more specific activity is to engage in community-based agroforestry schemes in order to try to maintain the natural environment for all users.

Vulnerability A set of conditions and processes resulting from physical, social, economical and environmental factors, which increase the susceptibility of a community to the impact of hazards.

Resilience The ability of a system, community or society exposed to hazards to resist, absorb, accommodate to and recover from the effects of a hazard in a timely and efficient manner.

Knowledge check 17

Explain how overgrazing leads to desertification.

Exam tip

Desertification is clearly linked to climate change. Revisit the themes of mitigation and adaptation in the previous section on climate change (page 19, The carbon cycle).

Exam tip

Desertification illustrates physical and human processes operating together, often involving feedback mechanisms. You could draw a diagram to help you clarify these in your head.

Case studies

You are required to have studied two case studies:

1 A case study of a hot desert environment, or a setting similar to it, to illustrate and analyse some of the key themes set out above and engage with field data. It may be that you have been to a hot desert area where processes associated with wind and water can be examined first hand. However, it is also recognised that a similar setting may be used that is *not* in a hot desert environment, such as an area of coastal dunes in the UK. In this latter scenario, it is most likely that you will have studied the action of wind (known as aeolian processes, see page 25. The key aspect is that you make use of data collection methods based on either moving sand or water in either area of study.

2 A case study of a setting in a desert at a local scale to illustrate and analyse key themes of desertification. The chosen area could be in any area of the world, such as the Sahel of Africa, the southwest USA or fringe areas of the Australian desert. You should look at its causes and impacts, and assess the implications of these impacts on sustainable development. You should also examine and evaluate the human responses to these causes and impacts with a view to assessing resilience, and the extent to which people can mitigate or adapt to the conditions there.

Exam tip

Exam questions on the second case study are likely to use one or more of the following words: sustainable, resilience, mitigation and adaptation. Make sure you understand these terms.

Summary

After studying this topic, you should be able to:

- explain how hot deserts operate as natural systems and describe their main characteristics, such as global distribution, climate, vegetation and soils
- explain the variety of ways in which deserts are caused
- discuss the range of geomorphological processes that operate in desert environments, including weathering, erosion, transportation and deposition by the actions of wind and water

- describe the variety of landscapes that have been created in desert environments over time, largely due to the actions of wind and water
- discuss the concept of desertification, its causes and impacts
- evaluate the impact of change in areas subject to desertification, such as climate change, and assess the implications for the futures of people living in these areas

■ Coastal systems and landscapes

Coasts as natural systems

Coasts are important and varying elements of the natural **landscape**. This diversity is a result of their history of sea-level change (some of which is taking place currently), the geological structures that lie behind them, the sediments that are available to make their beaches and the nature of waves and tides that mould them. There is a variety of processes (erosion, deposition and sea-level change) operating along coastlines. Coasts provide many opportunities for work in the field to examine how these processes operate and what they produce. They develop a variety of landforms which, when assembled, result in the varying landscapes that have developed along coastlines.

There is also a variety of strategies as to how coastlines should be protected and managed, if indeed they should be at all. As elsewhere in this subject, sustainability is a key criterion that should be considered.

Coastal environments are natural, open systems with inputs and outputs and flows between the two.

Inputs consist of:

- energy provided by waves, winds, tides and currents. This can be sporadically increased at times of storms and storm surges
- sediment provided within the system from the erosion of coastlines by waves, as well as that brought by rivers. Weathering and mass movement also contribute material from cliff faces
- changes in sea level — as sea levels rise with climate change, then more energy is exerted on a coastline

Outputs consist of:

- coastal landforms, both erosional and depositional
- accumulations of sediment above the tidal limit (dunes)
- loss of wave energy through processes such as refraction

Stores in the coastal system refer to the water in the sea, and sand/shingle on beaches. **Transfers** result from the actions of wind and waves. An example of a transfer mechanism is the process of longshore drift. (You should refer back to the section on **feedback** and **dynamic equilibrium** in the earlier part of this book, page 6.)

Systems and processes

Waves

Waves are caused by the wind blowing over the surface of the sea. As the wind drags over the surface of the water, friction causes a disturbance and forms waves. Waves at sea follow an orbital movement and objects on the water do not travel forward. The resultant up/down movement at sea is called the swell. However, when a wave reaches shallow water, the movement of the base of the wave is slowed by friction with the seabed, the wave spills forward as a breaker, moving objects forward with it. Water rushes up a beach as swash, before drawing back to the sea as backwash.

Wave energy is controlled by the:

- force of the wind and its direction
- duration of wind
- fetch — the longer the fetch, the more energy waves possess

Sustainability Meeting the needs of today without compromising the ability of future generations to meet their own needs.

Knowledge check 18

Define two key terms associated with coasts: fetch and tides.

It is common to classify waves as being either constructive or destructive:

- **Constructive** waves construct or build beaches and are usually the product of distant weather systems. They have longer wavelengths, lower height and are less frequent, at 6–8 per minute. Swash is greater than backwash, so they add to beach materials, giving rise to a gently sloping beach. The upper part of such a beach is marked by a series of small ridges called berms, each representing the highest point the waves have reached at a previous high tide.

- **Destructive** waves have a shorter wavelength, a greater height and are more frequent, at 10–14 per minute. The backwash is greater than the swash so that sediment is dragged offshore. This creates a steeper beach profile initially, though over time the beach will flatten as material is drawn backwards. Destructive waves also create shingle ridges at the back of a beach, known as storm beaches, as it is often local storms that create them.

On sandy beaches with large waves, the underwater part of the beach is often characterised by alternating shallow and deep sections, referred to as **ridges** and **runnels**. Sometimes these channels can either run parallel to the beach or at right angles to it. Strong currents can be present in these channels and when their flow speed is high, they can be dangerous.

High-energy coastlines are those in which wave power is strong for a greater part of the year, for example the western coast of the British Isles. The prevailing and dominant wind direction on these coasts is westerly and they face the direction of the longest fetch. The maximum-recorded wave height on western coasts is therefore greater than that on eastern coasts. Waves up to 30 m have been recorded on the west coast of Ireland.

Many estuaries, inlets and sheltered bays are **low-energy coastlines**, where wave heights are considerably lower. Here waves spread outwards and energy is dissipated, leading to the deposition of transported material. Enclosed seas also contain low-energy environments. The Baltic Sea contains some of the longest depositional landforms in the world because of its sheltered water and low tidal range.

Sediment sources and cells

Sediment sources include rivers, the seabed, erosion of the coastline, shell material and movement of materials along the coastline. **Sediment** (or **littoral**) **cells** are distinct areas of coast separated by deep water or headlands within which material is moved but the sediment inputs and outputs are in balance — a type of sediment budget. In the UK, the Department for Environment, Food and Rural Affairs (DEFRA) defines a sediment cell as 'a length of coastline and its associated nearshore area within which the movement of coarse sediment (sand and shingle) is largely self-contained. Interruptions to the movement of sand and shingle within one cell should not affect beaches in a neighbouring sediment cell'. In theory, they are regarded as closed cells (from which nothing is net gained or lost) but, in reality, finer material does move into neighbouring cells.

Eleven major sediment cells have been identified for England and Wales as basic units for coastal management. Sediment cell theory is a key component of Shoreline Management Plans, the authors of which decide future strategies for coastal management.

Exam tip

When asked to compare or contrast features, such as different types of wave, make sure you make clear comparative statements rather than separate statements.

Knowledge check 19

Describe the nature of strong currents known as rip currents.

Sediment budget The relationship between accretion and erosion, which can be used to predict the changing shape of a coastline over time.

Weathering

Physical weathering involves the breakdown of rocks into smaller fragments through mechanical processes, such as expansion and contraction mainly due to temperature change. Two types of physical breakdown that are common on coasts are frost shattering weathering and salt weathering. These are examples of sub-aerial weathering.

- **Frost shattering** (or **freeze–thaw**) is a form of physical weathering in rocks that contain crevices and **joints**, and where temperatures fluctuate around 0°C. Water enters the joints, and during cold nights freezes. As ice occupies around 9% more volume than water, it exerts pressure within the joint. This alternating freeze–thaw process slowly widens the joints, eventually causing bits to break off from the main body of rock. It leads to the formation of **scree** slopes.

- **Salt weathering** takes place when a rock becomes saturated with water containing salt, as in coastal environments. Some of the salt crystallises and begins to exert pressure on the rock, as the salt crystals are larger than the spaces in which they are being formed. As with frost shattering, the process repeats over time and causes the disintegration of rock.

Chemical weathering involves the decay or decomposition of rock in situ. It usually takes place in the presence of water, which acts as a dilute acid. The end products of chemical weathering are either soluble and are therefore removed in solution, or they have a different volume — usually bigger — than the mineral they replace. The rate of chemical weathering tends to increase with rising temperature and humidity levels, except in the action of carbonic acid (carbonation), where lower temperatures produce greater rates of weathering on limestones. Chemical weathering can also occur from the action of dilute acids resulting from both atmospheric pollution (sulfuric acid), and the decay of plants and animals (organic acids).

Weathering on a coastline is not only determined by the geology, but also by the climate. In high latitudes, freeze–thaw action (or frost shattering) is likely to be highly active along rocky coastlines.

Another means by which material reaches the sea is through **runoff** — the flow of water overland either as rills in small channels, or as rivers and streams.

Mass movement

The rate of mass movement depends on the degree of cohesion of the weathered material (regolith), the steepness of the slope down which the movement takes place and the amount of water contained in the material. A large amount of water adds weight to the mass, but, more importantly, lubricates the plane along which movement can take place.

There are various forms of mass movement that affect coastlines:
- **Creep:** the slow downhill movement of soil and other material. It tends to operate on slopes steeper than 60° and evidence of it is shown by small **terracettes** on a hillside.
- **Earthflows:** when weathered material becomes saturated, internal friction between the particles is reduced and stress can cause the debris to move under gravity. This can occur on slopes as gentle as 5° once mobile, but usually needs a slope of about 10° to initiate movement. Such flows are generally faster than creep.

Sub-aerial weathering
The collective name for weathering processes on the Earth's surface (literally at the base of the atmosphere).

Exam tip
Ensure that you know the difference between weathering and erosion processes.

Mass movement The downslope movement of weathered material under the influence of gravity.

Regolith The collective name for all of the material weathering produces.

- **Mudflows:** more rapid flows that can occur on relatively low slope angles compared with earthflows. They occur in areas that experience torrential rain falling on ground that has limited protection from vegetation cover. This allows the regolith to become saturated, increases the pore water pressure in the debris and reduces the frictional resistance between particles.

- **Rock falls:** when erosion is concentrated at the base of a cliff, it will become unstable and collapse into the sea. This is a common feature along the chalk cliffs of southern England.

- **Landslides (landslips):** occur when rocks and/or regolith have bedding planes or layers, and material in one plane/layer becomes very wet and over-lubricated. The added weight from the water will cause the plane/layer to slip downwards under gravity over the underlying layers.

- **Slumping:** saturated material moves suddenly, resulting in whole sections of cliffs moving down towards a beach. This happens particularly where softer material overlies strata that are far more resistant. The slip plane is often concave, producing a rotational movement.

Knowledge check 20

What are the key differences between a flow, a slide and a slump?

Erosion, transportation and deposition

Coastal erosion operates through a variety of processes:

- **Abrasion (corrasion):** the material that waves carry (their load) is used as ammunition to wear away rocks on a cliff or a wave-cut platform when repeatedly thrown or rubbed against these landforms. Where abrasion is targeted at specific areas, such as notches or caves, it is referred to as **quarrying**.

- **Attrition:** the process by which loose rocks are broken down into smaller and more rounded pebbles, which are then used in abrasion.

- **Hydraulic action:** where a wave breaking against rocks traps air into cracks in the rock under pressure, which is then released suddenly as the wave retreats. This causes stress in the rock, which develops more cracks, allowing the rock to break up more easily. Hydraulic action also includes **pounding** — the sheer weight and force of water pushing against a cliff face causing it to weaken.

- **Cavitation** occurs when air bubbles trapped in the fast-moving water collapse, causing shock waves to break against the rocks under the water. Repeated shocks of this nature are thought to weaken the rock.

Exam tip

Despite what it may say in the specification, solution is *not* a major form of erosion by the sea — it is too alkaline.

The rate of coastal erosion is governed by several factors:

- **Geology:** harder rock, for example granite, is more difficult to erode than softer material like boulder clay. Some more resistant rock can be eroded along joints or cracks, and limestones are prone to weathering. The structure and dip of rocks also affect erosion. If rocks dip inland or are horizontally bedded, steep cliffs form, whereas those rocks that dip seaward produce gentler slopes.

- **Coastal shape:** softer rock is eroded to form bays, with the harder rocks forming headlands. Wave refraction is concentrated on headlands, causing erosion here, whereas in the bays the waves spread over a wider area and their energy is dissipated, causing more deposition.

- **Wave steepness:** steeper, high-energy waves have more power to erode.

- **Wave breaking point:** waves that break at the foot of a cliff have higher energy to erode.

- **Fetch:** waves that have travelled a long distance have more energy.
- **The size and type of beaches:** beaches absorb some of the waves' energy and protect the coastline. Pebble beaches dissipate energy from waves through friction and percolation.
- **Human development:** sand and pebble extraction from beaches for use as building materials weakens the coast's protection and can lead to erosion. Sea walls, groynes and other coastal protection schemes may help to protect an area from erosion, but can also increase erosion further along the coast. Developments on the top of cliffs can increase runoff, and cause instability and cliff failure.

Material is **transported** along a coastline by:

- **swash and backwash:** sand and shingle are moved up the beach by swash and back down the beach by backwash
- **longshore (littoral) drift:** material is moved along the shoreline by waves that approach the shore at an angle. Swash moves sand and shingle up the beach at an angle but the backwash is at right angles to the beach. This results in material zigzagging its way along the beach according to the prevailing wave direction. Obstacles such as groynes interfere with this drift, and accumulation of sediment occurs on their windward sides, leading to entrapment of beach material. Deposition of this material also takes place in sheltered locations, such as where the coastline changes direction abruptly, creating sand spits.

Deposition occurs in low-energy environments, such as bays and estuaries. When sand is deposited on a beach — i.e. when the swash is more dominant than the backwash — and dries out, it can be blown by the wind further inland to form sand dunes at the back of the beach. In a river estuary, mud and silt can build up in sheltered water to create a salt marsh. Here the freshwater of the river meets the salt water of the sea, causing flocculation to occur and creating extensive areas of mudflats.

Coastal landscape development

Erosional landscapes

Cliffs, **headlands** and **bays** form when rocks of differing hardness are exposed together at a coastline. Tougher, more resistant rocks (such as granite and limestones) tend to form headlands with cliffs. Weaker rocks (such as clays and shales) are eroded to form sandy bays. In addition to these straightforward factors, the process of wave refraction is also important.

On an indented coastline, headlands and the offshore topography concentrate wave attack on that headland. Many headlands have a **wave-cut platform** between high and low tide, which can cause friction for the wave. However, due to headlands' solid nature they do not absorb energy, as a sandy beach would do, so waves can break at the foot of the cliff, causing maximum erosion. In a bay, waves have to travel further and a beach absorbs wave energy and reduces the power of the wave before it reaches the cliff. Where there is a wide, deep, sandy beach, waves may not even reach the backshore at all.

Knowledge check 21

Explain the process of 'wave refraction'.

Flocculation The process by which a river's load of clays and silts carried in suspension is deposited more easily on its meeting with sodium chloride in seawater.

Exam tip

Questions on this area of study will require an understanding of the distinctive nature of the processes in this environment. References to sequence and time are key here.

A cliff base will erode faster than the rock above it due to breaking waves in the tidal zone. A **wave-cut notch** develops here and rocks above may overhang. Gravity causes these overlying rocks to fall and the cliff line retreats, leaving the remnants of the rocks of the original cliff line as an almost macro-scale flat (less than 4°) base called a **wave-cut platform**. However, at a micro scale, most wave-cut platforms have a series of small terraces, micro-cliffs and deep rock pools, which can make clambering over them quite dangerous, especially if they are covered with seaweed.

At any point on a cliff coastline where there is a weakness, such as cracks, joints or along bedding planes, erosion can take place. Where waves open up a prolonged joint in the cliff face, they form a deep and steep-sided inlet, or geo. If this inlet is widened to become a small **cave**, a crack at the back of the cave may open up like a chimney to the surface, creating a hole at the top of the cliff above. This is often referred to as a **blow hole**.

Where caves are created on headlands and are eroded back into the headland, they sometimes meet a cave that is being similarly eroded on the other side of the headland. The back wall separating the caves weakens and eventually the sea pushes straight through, forming an **arch**. The sea is now able to splash under the arch, further weakening it until eventually the arch roof collapses, leaving the seaward side of it as a separate island called a **stack**. Over time the stack will be eroded to form a **stump**.

Depositional landscapes

Coastal deposition takes place on sheltered stretches of coast. Sediment that can no longer be transported along the seabed or is suspended in water is deposited to form features such as **beaches** and **spits**, which build up by accretion.

Beaches are commonly found in bays. There, wave refraction creates a low-energy environment that leads to deposition. Beaches are made of either sand, shingle or a mixture of both. This nature of the available sediment and the power of the waves influence the nature of deposition. Beaches are often classified as being either swash-aligned — where sediment is taken up and down the beach with little sideways transfer — or drift aligned, where sediment is transferred along a beach by longshore drift.

Beaches can be sub-divided into different zones indicating the position of aspects of the beach in relation to breaking waves and tidal ranges:

- **Offshore:** beyond the influence of breaking waves
- **Nearshore:** intertidal and within the breaker zone
- **Foreshore:** the surf zone
- **Backshore:** usually above the influence of normal wave patterns, marked at the lower end by berms, and may have a storm beach further up

Within the offshore and nearshore zones, a number of minor beach landforms can be found:

- **Ridges and runnels** (alternate raised sand bar and dip sections) run parallel to the shoreline and are caused by periods of strong backwash, sometimes associated with strong tides. They are often exposed at low tide, but are hidden at high tide, and can hence cause some difficulties for swimmers.

Exam tip

You are required to know examples of all of these landforms from the UK and other areas beyond the UK.

Exam tip

A good way to demonstrate that you know what each of these landforms looks like is to draw a sketch of it. Give it a go!

Knowledge check 22

Explain why wave-cut platforms tend to have a maximum width of about 0.5 km.

Accretion The growth of a natural feature by enlargement. In the case of coasts, sand spits grow by accretion, as do other landforms such as sand dunes.

Exam tip

You will be asked to describe landforms. Refer to size, shape, nature of sediments and field relationship (i.e. where the landform lies in relation to the landscape).

- **Beach cusps** are a series of small depressions that develop where beaches are made of sand and shingle. The sand is worn away more easily than the shingle and creates a semi-circular hollow in a miniature bay-like formation. Once created they self-perpetuate, especially on swash-aligned beaches.
- **Ripples** are micro-beach ridges parallel to the shoreline created by wave action in the foreshore zone on low-gradient beaches.

Spits are long, narrow stretches of sand or shingle that protrude into the sea. They result from materials being moved along the shore by longshore drift. This movement continues in the same direction when the coastline curves or where there is an estuary with a strong current that interrupts the movement of material and they project out into it. The end of the spit is often curved (creating a series of laterals) where waves are refracted around the spit end into the more sheltered water behind. This is also helped by the direction of the second-most dominant wind. If a spit joins the mainland at one end to an island at the other it is called a **tombolo**.

Where a spit has developed right across a bay because there are no strong currents to disturb the process, it creates a **bar** that then dams the water behind it, forming a lagoon. Bars also develop as a result of storms raking up pebbles, and this shingle left in ridges offshore creates a type of **barrier beach**.

Offshore bars are deposits of sand and shingle situated some distance from a coastline. They usually lie below the level of the sea, only appearing above the level of the water at low tide. There are two explanations as to where and how they form:

1 in shallow seas where the waves break some distance from the shore
2 where steep waves break on a beach, creating a strong backwash that carries material back down the beach, forming a ridge

When a bar begins to appear above the level of the sea for most of the time, it also becomes a barrier beach. In this case, a series of elongated islands stands above sea level, with a lagoon on the landward side and the ocean on the other.

Sand is often deposited by the sea under normal low-energy conditions. Wind may then move the sand to build **dunes** further up the beach, which in turn become colonised by stabilising plants — a process that results in a **psammosere**.

Estuarine environments

Sheltered river estuaries or the zones in the lee of spits are areas where there are extensive accumulations of silt and mud, aided by the process of flocculation and gentle tides. These are called **mudflats**. These inter-tidal areas will be colonised by vegetation and, as with sand dunes, a succession of plant types may develop over time, called a **halosere**. The resultant landform is called a **salt marsh**.

The initial plants of a halosere must be tolerant of both salt and regular inundation at high tide. These are called halophytes and include eelgrass and spartina grass. The latter has both a long root system and a mat of surface roots to hold the mud in place. These plants trap more mud and build up a soil for the next stage so that plants such as cord grass, sea lavender and sea aster can grow. As the mat of vegetation becomes more dense, the impact of the tidal currents reduces and humus levels increase, allowing reeds and rushes to grow next, and then eventually alder and willow.

Exam tip

You will be asked to describe landforms. Refer to size, shape, nature of sediments and field relationship (i.e. where the landform lies in relation to the landscape).

Exam tip

A good way to demonstrate that you know what each of these landforms looks like is to draw a sketch of it. Give it a go!

Psammosere The succession of plants that develops on a sand dune complex. Plants include sea rocket and lyme grass nearer the sea, with marram grass, fescue and gorse inland.

Salt marshes often have complex systems of waterways, known as creeks. In some extensive salt marsh areas hollows of trapped seawater form, which then evaporates, creating salt pans.

Landscapes of sea-level change

Changes in sea level take place over time due to sea temperatures being colder or warmer than the present, or relative changes in land levels.

Eustatic change may result from a fall in sea level due to a new glacial period, when water is held as ice. This explains why during previous glacial periods the English Channel was dry land. At the end of a glacial period, the ice on land melts and global sea levels rise again.

Isostatic change arises from changes in the relationship of land to sea. During a glacial period, as ice collects on the land, the extra weight presses down on the land causing it to sink and the sea level to rise. As the land ice melts the land begins to move back up to its original position (isostatic readjustment), and the sea level falls. This process is variable depending on the thickness of the original ice and the speed of its melting.

Tectonic processes associated with plate movement have also caused changes to sea level. By their nature, they tend to be quite localised.

In the last 10,000 years, a geological period called the Holocene — and especially between 10,000 and 6,000 years ago — saw the global sea level rise very quickly. It flooded the North Sea and English Channel, broke the link between Britain and Ireland and flooded many river valleys to give the distinctive indented coastline of southwest England and Ireland, known as rias (see below). This rise in sea level is known as the Flandrian transgression. Since then, sea levels have remained largely consistent.

The effect of a rise in sea level is to flood the coast, creating a coastline of **submergence**. Rising sea levels have flooded pre-existing valleys. **Rias** are drowned river valleys with long fingers of water stretching a long way inland, including the tributary valleys. They are widest and deepest nearer to the sea and get progressively narrower and shallower further inland. They are often winding, following the shape of the pre-existing river valley. Tidal changes will often reveal extensive areas of mudflats. **Fjords** are glaciated valleys that were drowned by the rising sea level at the ends of the Ice Age. They often have a shallower area at the mouth called a rock threshold, where the ice thinned as it reached the sea and hence lost its erosional power. Fjords have the typical steep-sided and deep cross profile associated with glacial troughs, and they can stretch many kilometres inland. The main channel is often straight, with right-angled tributary valleys.

Dalmatian coast is the name given to a drowned coastline where the main relief trends run parallel with the line of the coast. The ridges of upland produce elongated islands separated from the mainland by the flooded valley areas. The name originates from the Adriatic coast of Dalmatia in Croatia.

The effect of falling sea levels is to expose land normally covered by the sea, creating a coastline of **emergence**. Cliffs that are no longer being eroded become isolated from the sea, forming **relic cliffs**. 'Fossil' features such as former caves and stacks are left higher up from the coast on raised **marine platforms** compared with present-day features. Less dramatic residual features are **raised beaches**. These are common on the coast of western Scotland, where a series of raised sandy and pebble-ridden terraces can be found above the level of current sea levels.

Sea-level rise associated with climate change is an important issue for the future, with increases of several centimetres (or more) expected over the coming decades. This sea-level rise is due to:

- the thermal expansion of water as it becomes warmer
- more water being added to the oceans following the melting of freshwater glaciers and ice sheets, such as those in Greenland

Coastal management

Some geographers classify coastal defence strategies against risks of coastal flood and erosion into '**hard engineering**' methods and '**soft engineering**' methods. Hard engineering approaches (Table 1) have significant financial implications and severe environmental costs, some of which are unpredictable. They tend to be focused on areas of greatest need, particularly urban areas with high land values and population densities. Soft engineering (Table 2) is less costly and less environmentally damaging. For these reasons, it is often stated to be more sustainable. Many modern coastal management schemes combine elements that are both 'hard' and 'soft' in an integrated approach.

Table 1 Hard engineering methods

Strategy	Description	Commentary
Sea wall	A concrete or rock wall at the foot of a cliff or at the top of a beach. It usually has a curved face to reflect waves back out to sea.	Although often effective at the location where they are built, they deflect erosion further along the coast. They are expensive and have high maintenance costs.
Groynes	Timber or rock structures built at right angles to the coastline. They trap sediment being moved along the coast by longshore drift.	The beach created increases tourist potential, and gives protection to the land behind. However, this same process starves beaches further down the coast of sand and increases erosion there.
Rip-rap (rock armour)	Large, hard rocks dumped at the base of a cliff, or at the top of a beach. It forms a permeable barrier to breaking sea waves.	Relatively cheap, and easy to construct and maintain. The rocks used are often brought in from other areas and hence may not blend in.
Revetments	Wooden barriers, in a slat-like form, placed at the base of a cliff or top of a beach.	They are intrusive and very unnatural.
Gabions	Wire cages filled with small rocks that are built up to make walls. Often used to support small, weak cliffs.	They are relatively inexpensive. They look unsightly to begin with but as vegetation grows they blend in better. The metal cages rust and break easily.

Knowledge check 24

Make three lists: coastal landforms created by erosion; coastal landforms created by deposition; and coastal landforms resulting from sea-level change.

Exam tip

You will often be asked to 'evaluate' the effectiveness of management strategies. To do this, consider their advantages and disadvantages.

Table 2 Soft engineering methods

Strategy	Description	Commentary
Beach nourishment	The addition of sand or pebbles to an existing beach to make it higher or wider. The materials are usually dredged from the nearby seabed and spread, or 'sprayed', onto the beach.	A relatively cheap and easy process, and the materials used blend into the natural beach. However, it is a constant requirement as natural processes continue to do their work.
Dune regeneration	The planting of marram grass and other plants that bind the sand together. Areas are often fenced off to keep people off newly planted dunes.	Maintains the look of a natural coastline and provides important habitats. The process requires a lot of time to be effective.
Marsh creation	Low-lying coastal lands are allowed to be flooded by the sea. The area becomes a salt marsh.	It provides an effective buffer to the power of waves, creating a natural defence, and an opportunity for wildlife habitats. However, agricultural land is lost and landowners require compensation.

> **Exam tip**
>
> For many of the strategies of coastal management identified, have an example of where they have been employed.

Shoreline Management Plans (SMPs)

The UK government introduced Shoreline Management Plans (SMPs) in 1995. These are an approach to coastal management that involves all stakeholders making decisions about how coastal erosion and coastal flood risk should be managed. They aim to balance economic, social and environmental needs with pressures at the coast. There are 22 SMPs in England and Wales. In Scotland and Northern Ireland the devolved governments and local authorities are jointly responsible for coastal protection.

SMPs provide a large-scale assessment of the risks associated with coastal processes and present a policy framework to reduce these risks to people and the developed, historic and natural environment in a sustainable way. They determine the natural forces that are sculpting the shoreline and predict, so far as it is possible, the way in which it will be shaped into the future, defined as 100 years.

> **Knowledge check 25**
>
> Who manages SMPs and why is this important?

During the coastal defence planning process of SMPs, four coastal defence policies are often considered:

1 **Hold the line:** maintaining or upgrading the level of protection provided by defences.

2 **Advance the line:** building new defences seaward of the existing defence line.

3 **Managed realignment (or retreat):** allowing retreat of the shoreline, with management to control or limit movement.

4 **No active intervention (sometimes referred to as 'do nothing'):** a decision not to invest in providing or maintaining defences.

Case studies

You are required to have studied two case studies:

1 A case study of a coastal environment at a local scale to illustrate and analyse fundamental coastal processes and the landscapes that have been created by these processes. It is important that you engage with field data within this location. You should also consider how such areas can be managed in a sustainable way, as such locations are often popular with tourists and yet used by locals. Examples in the UK include the Yorkshire coast, such as Flamborough Head, or the Dorset Jurassic Coast.

2 A case study of a coastal environment beyond the UK to illustrate and analyse coasts as presenting risks and opportunities for human occupation and economic development. It is important that you consider how the people of this area have responded to the risks and opportunities within the key themes of resilience, mitigation and adaptation. Examples could include the delta coastline of the Netherlands or the coral reefs off northern Australia.

Exam tip

The exam questions on this case study are likely to use one or more of the following words: sustainable, resilience, mitigation and adaptation. Make sure you understand these terms.

Summary

After studying this topic, you should be able to:
- explain how coasts operate as natural systems, and be able to describe their main inputs (wind, waves and tides) and outputs (landforms)
- discuss the range of geomorphological processes that operate in coastal environments, including weathering, mass movement, erosion, transportation and deposition by the actions of wind and waves
- describe the variety of landscapes that have been created in coastal environments over time, largely due to the erosional and depositional actions of wind and waves
- explain the variety of mechanisms of sea-level change (eustatic, isostatic and climate change) and the amended landforms due to submergence and emergence
- discuss the concept of coastal management against flood and erosion, and the variety of approaches by which it can be addressed sustainably

■ Glacial systems and landscapes

Glaciers as natural systems

Glaciers and their immediate fringes create very distinctive **landscapes**, many of which remain visually stunning for thousands of years after their creation. Within the label of glacial, we also consider the landscapes created by meltwater from glaciers and ice sheets (**fluvioglacial**) and those created by very cold climates, even though glaciers may not have been involved in their creation (**periglacial**).

The most obvious **input** into the glacial system is snow. Other forms of precipitation can also fall on a glacier, adding to the mass when they freeze. This overall gain is

Exam tip

Once you have completed this option, be clear of the difference between glaciation, fluvioglaciation and periglaciation.

mass is called **accumulation**. A glacier also has rock fragments on its surface and embedded within and under the ice. This is also part of the mass of a glacier and is important, as it is the tool that carries out erosion. Snow and rock can fall onto the surface of a glacier as a result of avalanches or from weathering processes like frost shattering on a valley's sides. Solar energy is also an input into the system. Ice and snow have a very high albedo, but the sun's energy still leads to melting and evaporation, and the loss of mass.

A glacier's main **outputs** are through melting, sublimation and evaporation. Sublimation occurs when ice is converted to water vapour without melting. This can take place when it is cold and the atmospheric conditions are dry. Most of this loss is very seasonal, occurring during the summer. The collective term for this output is **ablation**. Material can also be transported away in meltwater streams. These are often a brown colour, reflecting their high load. Less obvious are the milky blue streams flowing from glaciers. They have this coloration due to the high content of very finely ground-down rock material called **rock flour**. Energy is also lost through reflection from snow and ice, and also radiation.

The main **store** of a glacial system is the glacier itself, though it will also include meltwater contained above, within and below it, as well as weathered and eroded material that forms part of the glacier's mass.

Transfers involve the movement of ice and water in the glacial environment. They are best considered under the term 'glacial budget' (see page 44).

The nature and distribution of cold environments

The global distribution of cold environments, present and past

Cold environments are typically those of the high latitudes, caused by their proximity to the North and South Poles. The equatorward boundary of cold climates is generally taken as the line where the mean temperature of the warmest month is no more than 10°C. This line broadly corresponds with the poleward limit of tree growth. In the northern hemisphere this isotherm swings south of the Arctic Circle into Alaska, Labrador, Greenland, Iceland and parts of northern Russia. In the southern hemisphere the only area of extensive cold is Antarctica. Other cold environments exist in the mountainous areas of the world — the Andes, Rockies, Alps, Himalayan mountains and Tibetan plateau.

To many people cold environments are synonymous with ice caps, sheets and glaciers. Around 10% of the world's surface is covered with ice but over the past few million years the extent of such ice has frequently been much larger than now, covering about one-third of the Earth's surface. Glaciers are important both because they cover substantial parts of the Earth's surface and because they have left their imprint on the land they formerly covered.

Albedo The amount of incoming solar radiation that is reflected by the Earth's surface and atmosphere. Fresh snow and ice have an albedo as high as 90%.

Knowledge check 26

In the context of glaciers, explain the difference between accumulation and ablation.

Knowledge check 27

Distinguish between the following types of cold environment: glacial, periglacial and alpine.

During the last ice age, 18,000 years ago, much of the northern hemisphere was covered in ice and glaciers. These covered nearly all of Canada, much of northern Asia and Europe, and extended well into the USA — in total, 30% of the land surface.

Current glaciated areas include the ice cap of Antarctica and northern Greenland, northern Russia, northern Scandinavia, northern Canada, and the high mountain areas of the high Andes, Rockies, Himalayas and Alps. Past glacial and periglacial environments include the UK and most of central Western Europe, where lower latitudes and altitudes show the imprint of the last (**Pleistocene**) glaciations. At the last glacial maximum, about 21,000 years ago, ice sheets covered vast swathes of Canada, northern Britain and Scandinavia.

Physical characteristics of cold environments

Cold environments can be subdivided into two types: ice caps and tundra.

Ice cap environments

Here average temperatures of all months are below freezing (0°C) so that vegetation growth is impossible, and a permanent ice and snow cover prevails. At the South Pole the warmest month (December) has a temperature of –28°C and the three coldest months (July, August and September) an average temperature of –59°C. Thermometers here have been known to record temperatures of –90°C. Precipitation is scant in amount as the low temperature, low humidity levels and the extreme stability of the air all inhibit snowfall. Precipitation levels are recorded as less than 150 mm per year.

Tundra

This is the name given to a climatic and vegetation type that can be found in the most northerly parts of North America and Eurasia (north of 65°N). The main features of the climate are:

- long and bitterly cold winters with temperatures averaging –20°C
- brief, mild summers with temperatures rarely being above 5°C
- a large temperature range of over 20°C
- low amounts of precipitation, less than 300 mm, most of which falls as snow
- strong winds blowing the dry, powdery snow in blizzards, and creating a high wind-chill factor.

Although the winters are severe and the sea regularly freezes, temperatures do not fall as low as areas further inland due to the moderating effect of the sea. The cold temperatures are due to the short hours of winter daylight in such areas, and although the daylight hours are longer in summer, the angle of inclination of the sun is low. High pressure with dry descending air dominates these areas. In summer, some depressions do penetrate, giving some precipitation.

Tundra is associated with large areas of frozen soil — **permafrost** (see periglaciation below). Due to the climate and the soils, these areas have a distinctive vegetation type, which features:

- dwarf species such as cotton grass, mosses and lichens growing close to the ground, often in a cushion-like form. This is an adaptation to the strong winds that blow in the area

Pleistocene A geological time period stretching from 2 million years BP (before present) to 10,000 years BP. It was characterised by a series of alternating cold phases (glacials) and warm phases (interglacials), collectively known as the ice ages.

Wind-chill The accentuation of cold temperatures when accompanied by high wind speeds. For example, an air temperature of 0°C with a wind speed of 6 ms^{-1}. has a chill factor of –10°C.

- dwarf willows and stunted birch trees in sheltered places
- plants that have long, dormant periods during the cold, dark winters, but grow rapidly in the summer when daylight hours are long. 'Bloom mats' of anemones, Arctic poppies and saxifrages burst into life, providing a mass of colour
- many plants that have small leaves to limit transpiration
- plants that are shallow rooted because the soil is permanently frozen at a shallow depth

Systems and processes

Glacial budgets

The **glacial budget** is the relationship between accumulation and ablation:

- If accumulation is greater than ablation, the glacier gains mass and the glacier snout advances.
- If ablation is greater than accumulation, the glacier loses mass and the glacier snout retreats.
- If accumulation is equal to ablation, the snout of the glacier remains in the same place — it is stationary.
- An **equilibrium line** can be shown on a glacier where at any one time the amount of accumulation equals that of ablation. This is also known as the **firn line**.
- Hence there are spatial variations in a glacial budget. Most of the accumulation occurs at the upper end of a glacier. At higher altitudes, where it is much colder, this exceeds losses due to ablation. There is therefore **net accumulation** and this part of a glacier can be called the **zone of accumulation**. At the lower end of a glacier the reverse is the case: ablation at warmer, lower altitudes exceeds accumulation, so there is **net ablation** and this part of a glacier is called the **zone of ablation**.

Note that there are usually seasonal changes in the glacial budget. Glaciers usually have a positive balance in winter, when more snow accumulates, and a negative balance in summer, when there is more ablation due to higher temperatures.

Historical patterns of ice advance and retreat can be examined for any one of the world's glaciers, and each glacier can have a different pattern of advance and retreat. You can track most of the world's glaciers using the following link: www.grid.unep.ch/glaciers/

Types of glacier

The largest glaciers are called **ice sheets**. These have a flattened dome-like cross section, are hundreds of kilometres in width and cover an area of more than $50,000\,km^2$. The most well-known ice sheets, because of their size, are those of Antarctica and Greenland. During the Pleistocene there were also two other enormous ice sheets, one over Scandinavia and Great Britain, and another over much of North America. Smaller domes, less than $50,000\,km^2$, are called **ice caps**. An **ice shelf** is an area of thick floating ice sheet attached to a coastline, whereas an **ice field** is a relatively flat and extensive mass of ice.

Exam tip

Note that glaciers *do not* retreat — it is their snout that retreats. Ice continues to move downhill but melts at a faster rate.

Exam tip

It is important that you understand fully the interplay between accumulation and ablation, including the seasonal variations in each.

The other main type of glacier is the one that occupies a valley or lowland basin. These can further be divided into the following:

- **Valley glacier:** body of ice that moves down a valley under the influence of gravity and is bounded by rock walls on either side.
- **Cirque glacier:** small body of ice that occupies an armchair-shaped hollow in the mountains, which has been cut into bedrock.
- **Diffluent glacier:** valley glacier that diverges from a main glacier and crosses a drainage divide through a diffluence col.
- **Piedmont glacier:** glacier that leaves confining rock walls and then spreads out to form an expanded glacier at the foot of the mountain valley.

An alternative classification of glaciers is according to their temperature:

- **Warm-based/temperate/alpine glaciers:** these are found in areas where temperatures are high enough to cause some summer melting. This allows the glaciers to advance in winter and retreat in summer. The meltwater reduces friction and lubricates the base, allowing the glacier to move more quickly. This allows more material to be transported and therefore more erosion and deposition occurs with these types of glacier.
- **Cold-based/polar glaciers:** these are found in areas where the temperature rarely rises above freezing so there is little melting or movement and much less erosion transport and deposition.

Geomorphological processes

Glacier formation

When snow crystals fall they have an open, feather-like appearance and a low density. However, if the snow crystals are compacted by the weight of overlying snow, or if they are partially melted, they are converted to a mass of partially consolidated ice crystals with interconnected air spaces between them. Such material is called firn, or neve. As this process develops further it becomes more dense, until most of the air spaces are eliminated and pure ice develops.

Most glaciers are composed of the ice produced from snow that has been modified in this way, but on some glaciers refrozen meltwater can make up much of the mass. Eventually sufficient ice may accumulate so that, under the influence of gravity, the mass begins to move and thereby becomes a glacier.

Glacial movement

Glaciers are generally not static. Most glaciers move slowly — less than 50 m a year. However, some glaciers do surge and may for short periods of time reach speeds of 5 m per hour, sometimes more.

The movement of a glacier takes place in three main ways: by sliding over bedrock; by alternate compression and extension of the ice mass in response to changes in the bedrock surface below the ice; and by internal deformation (creep) of the ice.

Basal sliding is linked to the pressure melting of ice that takes place along its base. Ice normally forms from water at a temperature of 0°C, but the temperature at which water melts is reduced under pressure. As a glacier moves it will exert pressure and therefore some melting may take place at its base. A thin film of water may then exist

> **Knowledge check 28**
> Compare the typical dimensions of valley glaciers and cirque glaciers.

> **Knowledge check 29**
> Describe the variations of movement within a glacier both in plan view and with depth.

between the glacier and the bedrock. This film reduces friction — acts as a lubricant — and allows the glacier to slide. Such movement is more likely to occur in warm-based glaciers, where temperatures are close to their melting point. Linked to this process is the concept of **regelation**. This is the refreezing of water under a glacier when pressure is diminished in the lee of an obstruction. Pressure melting takes place on the upstream side, creating a film of water, and regelation takes place on the downstream side, and the combination of both processes allows a temperate glacier to slide downhill.

Compressional and extensional flow take place when ice cannot deform sufficiently quickly to stresses beneath the ice. For example, there may be rapid changes in the gradient of the bedrock — steep sections followed by more gentle sections. As a result, ice fractures and movement take place along a series of planes. Tensional fractures (caused by extension) create crevasses in the upper layers of the glacier. Shear fractures (caused by compression) result in the thrusting up of ice blocks along faults. The outcome of this form of flow is a very irregular surface to the glacier.

Internal deformation is much smaller in scale and takes place in all glaciers, but it is the most common form of movement in cold-based glaciers. It takes place where the ice crystals set themselves in line with the movement of the glacier and slide past each other. Movement can occur along lines of weakness called cleavage planes. This is sometimes called **laminar flow**, where it can be linked to layers of annual accumulation.

Glacial weathering and erosion

A glacier performs erosion as it has the materials to do so provided to it. A glacier receives material through a combination of weathering and rock fall. Weathered material comes from frost action and nivation.

Frost action occurs where diurnal temperatures hover around freezing point (above during the day and below at night). Water in cracks in the rocks freezes at night, widening the crack by expansion (usually by 9%). In the day, more water collects in the crack, which again freezes, widening it further. The rock then shatters into angular fragments that fall onto the ice and are used as the tools for erosion by abrasion, or they form mounds at the base of the slope, forming scree. At a much smaller scale, when water freezes in a rock, the ice attracts very small particles of water that have not frozen from the adjoining pores and capillaries. Nuclei of ice crystal growth are thus established, and some believe this form of ice crystal growth to be a more potent form of rock shattering.

Nivation is where physical and chemical weathering under a patch of snow, due to diurnal and seasonal temperature changes, causes the rock to disintegrate. Subsequent meltwater washes away the weathered material, and a small hollow, called a nivation hollow, deepens.

Although glacier ice itself does not cause marked erosion of a rock surface, when it carries the debris that has either fallen onto it, or washed into it, abrasion occurs. Sharp rock fragments carried by the ice embed themselves in the base and sides of the ice. This is used to grind down the bedrock like sand paper, making it smooth and

Knowledge check 30

Glaciers can also move by surges. Describe and explain such movements of ice.

Exam tip

Although weathering and erosion operate together, in an exam context you need to be clear on the differences between the two sets of processes.

wearing it away. It can leave scratches on the rock in the direction of ice movement, called **striations**. Much of the debris in glaciers is ground down to a fine mixture of silts and clays, known as rock flour.

Glaciers can also cause erosion by **plucking**. If the bedrock beneath the glacier has been weathered in periglacial times, or if the rock is full of joints (well-jointed), the glacier can detach large particles of rock and take them with it. As this process continues, some of the underlying joints in the rock may open up still more as the glacier removes the overburden of dense rock above them — a process known as **pressure release**.

Glaciated landscape development

Landforms and landscapes due to glacial erosion

As debris-laden ice grinds and plucks away the surface over which it moves, characteristic landforms are produced, which give a distinctive character to glacial landscapes. One of the most impressive landforms is the **cirque (corrie, cwm)**. This is an armchair-shaped, steep-sided hollow at the head of a glaciated valley. Cirques are often N or NE facing in the northern hemisphere, as this direction will receive the least sunlight, and are in the lee of prevailing winds, causing the snow to accumulate longer. Their size varies but they are often around 0.5 km in diameter with a back wall up to 1,000 m in height.

The hollow is deepened initially by nivation and then, as the ice accumulates, by rotational slide. This enhances the abrasion process due to increased compression in the base of the hollow. Ice pulls away from the back wall, plucking rocks already loosened by freeze–thaw weathering. This creates a **bergschrund** — a crevasse that forms where the ice starts to pull away from the back wall of the cirque. Material falls down this crack and is embedded in the base of the ice, which is then used to abrade and help deepen the hollow. Where the pressure and erosion are lower at the front edge of the hollow there is deposition of moraine at the lip of the cirque, which often allows water to accumulate behind it in a post-glacial period, forming a **tarn**.

> **Exam tip**
>
> You are required to know examples of all of these landforms from the UK and other areas beyond the UK.

As cirques develop they eat back into the mountain mass in which they have developed. When several cirques lie close to one another, the divide separating them may become progressively narrowed until it is reduced to a narrow precipitous ridge called an **arête**. Should the glaciers continue to erode away at the mountain from all sides, the result is the formation of a **pyramidal peak** or **horn**.

Glacial valleys (or **troughs**) develop where glaciers flow into pre-existing river valleys. They widen and deepen the original valley, making it steep-sided with a wide, flat base. Glaciers tend to straighten the valley, cutting off spurs and leaving cliffs called **truncated spurs**. At the upper end of the valley, where the glacier has entered the valley from the corries above, there is often a steep wall called the **trough end**. Glacial flow is uneven and where compressing flow has over-deepened a section of the valley, or where there is softer rock that is more easily eroded, it forms a rock basin, which often fills with water post-glacially, forming a **ribbon lake**. A **hanging valley** forms where ice in a tributary valley cannot erode effectively because its movement is blocked by ice in the main valley (or it may be smaller in size). It therefore remains higher and smaller than the main glacial trough (hanging above it). Post-glacially, a hanging valley may be identified as a waterfall into the main valley below.

Roches moutonnées are isolated rocks, generally 5–30 m in height, along the base of a glaciated valley. They are characteristically smooth on one side and jagged on another. They form where a glacier moves over a band of harder rock and it smoothes the upstream side by abrasion, often leaving **striations** to show the direction of movement. As the glacier spills over the top of the obstruction it removes the loose rocks by plucking, leaving a jagged edge on the downstream side.

A number of high-latitude coastlines, such as those in Norway, western Scotland and South Island, New Zealand, are flanked by narrow troughs called **fjords**. These are submerged glacial valleys. Some of them are extremely deep (over 1,000 m in depth) and bear witness to the substantial erosional powers of glaciers. These great depths are usually separated from the open oceans by a sill of solid rock, or **threshold**, where the depth is only 200 m. This is thought to be due to the fact that as the glaciers met the sea their erosional power became much less.

Glacial transportation and deposition

The debris transported by ice may be divided into three main categories:

1 **Englacial debris:** material carried within a glacier.
2 **Supraglacial debris:** material carried on a glacier surface.
3 **Subglacial debris:** material carried at the base of a glacier.

Deposition of this debris is a complex process. In general, deposition occurs when the carrying capacity of the ice decreases because the gradient decreases, the velocity of the glacier decreases or more material becomes available than can be transported. The material deposited directly from the ice, which is smeared on to the landscape, is called **till** (or **ground moraine**). It consists of a wide range of grain sizes, from clays to stones, and is often called **boulder clay**. It also possesses very little stratification, frequently contains far-travelled erratic material and tends to have clasts (or stones) with edges and corners blunted by abrasion. Such till often has its larger particles showing an orientation, or alignment, to the movement of the glacier.

Deposited till may form a variety of distinctive landforms. Ridges of glacially deposited material occur along the end (terminal) margins of glaciers and ice sheets — these are collectively called **end moraines**. A series of end moraines may be traced across a lowland region for hundreds of kilometres, and may be up to 100 m in height.

Other moraines are associated with glaciers:

- **Lateral moraine:** moraine produced at the side of the glacier, having rolled there after falling on top of the ice or having fallen onto the side from frost shattering of the valley sides.
- **Medial moraine:** found at the confluence of two glaciers where two lateral moraines merge to form a deposit down the middle of the main glacier.

Drumlins

A **drumlin** is an oval-shaped hill, often of unsorted till, shaped like an egg half buried along its long axis. It can be up to 50 m in height and around 1,000 m long.

Its long axis runs parallel to the direction of ice flow, with its steeper 'stoss' end pointing up-ice and its gently sloping 'lee' end pointing down-ice. Drumlins can be either rock-moulded or composed entirely of glacial or pre-existing sediments. They are often found in groups or swarms where the landscape is described as 'a basket of eggs' topography.

Geographers still do not know exactly how drumlins are formed beneath the ice. The most widely accepted theory suggests that different sediments, with different strengths, behave in different ways when the overlying ice applies stress to them. This leads to zones of stronger sediments able to resist deformation, which then act as cores that develop into drumlins.

Erratics

Erratics are rocks that have been transported by a glacier or an ice sheet and deposited in an area of different geology from that of its source. Therefore, they can be used as evidence to indicate the direction of ice movement. Rocks from Ailsa Craig, in the Firth of Clyde, Scotland, have been found in southwest Lancashire, England, 240 km from their source.

Fluvioglacial processes and landforms

Meltwater flowing subglacially is often at high pressure. It can erode spectacular subglacial channel networks. The pressure normally forces meltwater towards the snout of the glacier, so subglacial channels can be used to trace the direction of ice flow. Lateral meltwater channels form when streams flow along the sides of the glacier or ice sheet, often following the contours of the slope, perched on the valley sides in parallel series. A number of landforms are found in a fluvioglacial landscape.

Eskers

These are the infillings of ice-walled subglacial, englacial or supraglacial river channels. When the ice melts away it leaves behind an elongate, sinuous ridge composed of sand and gravel, which can be up to 100 m in height and extend for tens of kilometres. Eskers may form single channels or branching networks. Eskers formed subglacially often have up-down profiles that reflect the shape of the ice/surface slope. Supraglacial and englacial eskers often have beaded (or flattened) shapes, reflecting the slow melt-out of underlying ice.

Kames

These are features produced at the margins of a glacier or ice sheet. They consist of irregular undulating mounds of bedded sands and gravels that are essentially a group of alluvial cones or deltas deposited unevenly along the edges of a stagnant or decaying ice sheet.

Proglacial meltwater channels

These are eroded by streams draining away from the ice margin. These channels follow the slope of the land and are often wide and deep. They are also often associated with lakes dammed by ice, that then spill over.

These meltwaters can create landforms known as **outwash plains (sandurs)** and **ice contact deltas**. Sandurs are large areas of often sandy and pebbly material washed

> **Exam tip**
>
> Glacial deposits are generally unsorted, with a mixture of sediment sizes and shapes. They tend to create a hummocky landscape.

> **Fluvioglaciation** The action of meltwater from glaciers and ice sheets.

> **Exam tip**
>
> It is important to appreciate the differing field relationships of these features: eskers are formed within a glacier, kames at the sides and proglacial means 'beyond the glacier'.

out of the ice by glacial meltwater streams. These streams carry a lot of debris in braided channels, which leave the largest deposits nearest the ice front and carry finer particles further across the plain. The deposits are also affected by seasonal melting and appear in layers, showing the larger deposits occurring each spring with the increased meltwater. These deposits are less angular than those left by the ice itself, and are sorted by the water leaving deposits over a wide area. During their creation, residual blocks of ice are often left behind by the retreating glacier. These melt over time, leaving small, often circular lakes, known as **kettle lakes**.

Periglacial processes and landforms

Examples of periglacial environments include those in high mountain regions such as the Rocky Mountains of North America, the interior plateaux of central Asia, and the tundra of Canada, Alaska and Russia. Periglacial environments are associated with **permafrost**. This is the permanently frozen subsoil lying beneath the surface of periglacial areas, which underlies approximately 20–25% of the Earth's surface. Permafrost can be continuous, discontinuous or sporadic, depending on the size of the area that it covers. The active layer lies above the permanently frozen area and temperatures in this layer rise above freezing in the summer. The active layer varies in thickness, but is usually between 5 m and 15 m in depth. One factor that can influence the depth of the permafrost is the amount of vegetation cover — it decreases with increasing levels of vegetation.

Several landforms are characteristic of periglacial areas. These are outlined below.

Pingos

These are domed mounds of layered sediments with a core of ice and are usually up to 100 m in diameter, though they can be as wide as 2 km. Pingos can be open or closed. **Open pingos** can be seen in the Canadian Arctic and Siberia and are more likely to be found in clusters. **Closed pingos** are less common and are found in Alaska and Greenland.

Whether a pingo is open or closed depends on its formation. Open pingos are formed in areas of discontinuous permafrost when groundwater from small areas of talik moves upwards because of an increase in hydraulic pressure as the active layer refreezes in winter. As the water freezes near the surface, it causes the ground surface to dome upwards, forming the characteristic shape of a pingo. Closed pingos are formed in areas of continuous permafrost when water pushed down from frozen lake sediments accumulates at depth, freezes and then expands again, causing the ground surface to dome.

Some pingos have a small depression at their top, resulting from localised melting there. Over time a pingo may collapse and fracture, leaving a ring-shaped scar in the landscape.

Patterned ground

This is one of the most common periglacial landforms, and can be found in Alaska, Greenland and northern Canada. Patterned ground can take the form of ice-wedge polygons or stone polygons.

Exam tip

Fluvioglacial deposits are generally sorted, often dominated by rounded or sub-angular pebbles, gravels and sands. The larger deposits are often orientated downstream.

Periglaciation The processes and landforms on the fringe of ice sheets and glaciers, as well as those areas where freezing conditions dominate.

Active layer The upper layer of the soil in permafrost areas where there is seasonal thawing.

Knowledge check 32

Distinguish between continuous, discontinuous and sporadic permafrost.

Talik A localised area where the permafrost has thawed.

Ice-wedge polygons

Ice-wedge polygons are generally 20–30 m across and are formed in areas of continuous permafrost by the effect of the ground freezing in the winter and thawing in the summer.

When temperatures fall in winter, water in fissures in the active layer freezes and therefore expands, pushing the ground apart. The wedge shape is maintained as water further down from the surface, over 3 m down, is cooler and remains frozen. Indeed, during the spring and summer more water and fine sediment flow down to this zone to refreeze, thereby perpetuating the process. The annual repetition of this process causes the gradual formation of marked ice-wedges, which develop in polygonal patterns.

Stone polygons and stripes

Stone polygons are smaller than ice-wedge polygons — less than 10 m across — and occur in both permafrost areas and high mountain areas. These are also the result of freezing and thawing, being generated by frozen soil and the development and expansion of small ice lenses under the surface (collectively known as **frost heave**):

- During the winter soils freeze downwards from the surface, reaching and 'grabbing' individual stones, which are pulled upwards by the vertical expansion of the frozen soil above them. The empty space left beneath is filled with loose unfrozen soil, so the stone is prevented from moving back when the soil below freezes.
- Stones are also pushed towards the surface due to the pressure of small ice lenses growing beneath them. This occurs because stones have a lower specific heat and become colder more quickly than fine-grained soil, so ice will form first directly beneath them, pushing them upwards. Stones also warm up more quickly so in the summer thaw, the ice in the soil beneath the stone melts first, allowing wet sediment to slump to fill the space beneath it. This prevents it from sinking back into its original position.

The net outcome of both of these processes is that stones move upwards in the talik. As the stones collect on the surface of the ground the larger stones are pushed towards the edge of the pile by ground expansion and gravity, and smaller stones, sands and silt are left in the middle, resulting in a polygonal shape. When polygons develop on a slope, the shapes become elongated and a sorted stone stripe is formed.

Solifluction lobes

In areas where there is discontinuous permafrost the subsoil does not thaw in summer so surface meltwater cannot percolate into the rock beneath. The soil becomes saturated and starts to flow downhill on even the gentlest of slopes (as low as 1°) — a form of **mass movement**. Reduced friction between particles due to the water content and a lack of vegetation to hold the soil together allows the soil to move, leaving tongue-like protrusions called **solifluction lobes**. Where the slope is steeper on the valley sides, **solifluction terracettes** form.

Other periglacial landforms

In some areas **blockfields (felsenmeer)** occur. Here the larger angular blocks of stone released by frost shattering on a cliff slowly spread across flat plains below, forming a field or sea of isolated rocks.

> **Exam tip**
>
> Be clear of the differences between the various types of polygon. Ice-wedge polygons are found in stone-free soils such as former lake beds; stone polygons in stone-rich soils.

Thermokarst refers to irregular hummocky terrain often studded with small water-filled depressions created by the melting of ground ice. Its development is the result of the disruption of the permafrost and an increase in the depth of the active layer. It is thought that the creation of thermokarst may increase substantially with climate change, and cause many buildings and features such as roads, bridges and pipelines to become much more unstable.

Human impacts on glaciated landscapes

The concept of environmental fragility

Fragile environments are those where any disruption to the ecosystem, however slight, can have serious short-term and longer-term effects. The tundra is considered to be a fragile environment because of its climate and limited biological productivity. The slow rate of plant growth means that any disruption to the natural equilibrium of the ecosystem can take a long time to be corrected, if at all. The low productivity of the area together with limited biodiversity mean that plants are very specialised and disruption causes difficulty when it comes to plant regeneration. In such circumstances, species, both plant and animal, have great difficulty in adapting to a changed environment.

Traditional activities in cold environments

The traditional economic activity of the indigenous population of the tundra was hunting and fishing. In the north of North America, the main activity of the **Inuit** was hunting seals, which provided them with meat, oil and skins. Fishing (including whale hunting) was also a major activity. The number of Inuit was always small in terms of the vast area in which they lived so that very little pressure was placed on the environment, which remained relatively undisturbed.

In the north of Europe, the **Sami** people of northern Scandinavia followed the seasonal movements of the herds of wild reindeer that provided them with most of the food and materials they needed. Fishing was used to supplement their diet. Reindeer spend the winter period in the boreal forests living off the tree mosses, lichen and bark. They then moved back into the tundra during the summer, a migration often involving distances of 300–400 km. Like the Inuit, the Sami lived in an environment that provided all that they needed but which could only support a low-density population. The ways of life that they adopted were totally sustainable.

However, in recent years their way of life has begun to change. Sami art objects and handicrafts are sold to tourists. In the winter pastures, lichen from old trees is an essential part of the reindeer diet, but due to modern forestry it is in short supply. Sometimes the reindeer are kept in enclosures and are fed hay and other commercial fodder during the winter. It is no longer a nomadic life, and many Sami now live in settled areas.

Modern activities in cold environments

The Alps is an example of a high-altitude cold environment. There are often conflicts between tourists and farming, and even different tourist groups in this area. Skiing and snowboarding off piste can set off avalanches, and tourist hotels, cable cars and ski lifts are said by some to ruin the landscape. The development of mountain sports and tourism in these remote areas has improved the transport network and economy by developing jobs in the tourist industry, which has kept more people living in the

> **Knowledge check 33**
>
> Make four lists: glacial landforms created by erosion; glacial landforms created by deposition; fluvioglacial landforms; and periglacial landforms.

Biodiversity The variety of different forms of life within an ecosystem.

area and prevented further rural depopulation. However, many people feel that the fragile environment has been damaged, with the loss of rare plants like orchids and birds such as the golden eagle.

Alaska has vast reserves of oil and gas. The reserves were first discovered around Prudhoe Bay in the north, which was inaccessible to ships due to pack ice. A pipeline needed to be built to the ice-free port of Valdez on the south coast of Alaska. This was constructed on insulated legs to prevent the warmed oil (warmed, so that the oil does not freeze in such low temperatures) from melting the permafrost and causing the pipeline to sink and break, and was raised on stilts to avoid blocking the migration route of the caribou. It also zigzags to prevent rupturing by ground movement due to either frost heave or earthquakes.

Modern buildings are insulated or raised above the ground with piles driven into the permafrost to prevent the ground warming up and developing thermokarst. Utilidors are used, which are insulated boxes, elevated above the ground, which carry water supplies, heating pipes and sewerage between buildings, and are designed not to melt the permafrost.

The impact of climate change

As climate change has increased, scientists have already measured a retreat in the permafrost zone. This is particularly acute in northwestern Canada and in Siberia. Increases in temperature of only 1°C have led to the trebling of the thaw rate in parts of central Canada. The immediate impacts of melting can be seen in the Arctic communities — buildings become undermined, roads subside unevenly and crack, and the supports holding pipelines can shift and even crack the pipelines. Given the world's dependence on oil and gas, the threat of pipelines having to be shut down is alarming. Another major concern regarding the melting of permafrost is the release of organic carbon. The soils of the permafrost are normally crammed with un-degraded, well-preserved organic matter in the form of leaves, twigs, roots etc. This is an enormous store of carbon, kept inert by being frozen in the ground. However, if the ground was to melt and the organic matter start to rot, carbon would be released as either carbon dioxide or methane, creating more greenhouse gases. This will melt more permafrost and so on, in a worsening positive feedback cycle.

The Arctic is estimated to contain about 900 Gt of carbon. Humans emit about 9 Gt of carbon from fossil fuels and deforestation every year. Hence, it would only take the release of 1% of carbon in Arctic permafrost soils to effectively double our emissions of greenhouse gases, especially methane. Although permafrost may seem an irrelevance to those of us in the temperate latitudes, it is clear that changes in it may have significant consequential effects on us through its impact of greenhouse gases.

Management of cold environments

The Arctic Council provides an important forum for the indigenous peoples of the Arctic to advance their agendas, including matters relating to the use of resources. This international convention provides a basis for groups to go back to their home countries and demand the rights specified by it.

Exam tip

You should focus on the *differences* between traditional activities and modern activities.

Exam tip

Revisit the section on feedback mechanisms in the earlier part of this book (page 6). Be sure you understand how they operate.

Knowledge check 34

Explain how other negative feedback cycles could actually reduce carbon emissions from melting permafrost.

Content Guidance

In Alaska, Native Corporations owns about 12% of the land, and the US federal government has a preference for subsistence regarding the use of resources located on the 60% of the land remaining under federal jurisdiction. The US federal government enables indigenous rights to use marine mammals and has established community development quotas to ensure that local groups have a stake in the region's marine fisheries.

In Greenland, there is no system of private property or well-defined rights to land and natural resources, but the indigenous-controlled Greenland Home Rule has authority to make most decisions about the use of terrestrial and marine living resources.

In Scandinavia, national institutions have traditionally regulated natural resources, and local stakeholders have had greater difficulty establishing their rights. An ongoing struggle to secure the Sami rights to land and natural resources has met with limited success. This is especially the case in Sweden, where the courts have generally denied claims to indigenous rights despite state recognition of these rights a century ago.

Oil, gas and mineral developments have generally provided few long-term jobs for local residents. In North America, however, where local governments and land claims organisations provide an institutional framework for mitigation and compensation, extractive industries have provided substantial cash amounts to local communities.

Knowledge check 35

What are the difficulties facing the peoples of the Arctic in their attempts to manage their environments at present and in the future?

Exam tip

Exam questions on the second case study below are likely to use one or more of the following words: sustainable, resilience, mitigation and adaptation. Make sure you understand these terms.

Case studies

You are required to have studied two case studies:

1 A case study of a glacial environment at a local scale to illustrate and analyse fundamental glacial processes and the landscapes that have been created by these processes. It is important that you engage with field data within this location. Examples in the UK include the Lake District, Snowdonia or the Highlands and isles of Scotland, such as the Cairngorms or the Isle of Arran.

2 A case study of a contrasting landscape beyond the UK to illustrate and analyse how it presents challenges and opportunities for human occupation and development. You should also use this case study to evaluate human responses to the environment in terms of resilience, mitigation and adaptation. Examples could include the exploitation of oil and gas on the fringes of the Arctic Ocean, or an examination of the future of ski resorts in the Alps.

Summary

After studying this topic, you should be able to:

- understand how glaciers and processes in cold environments operate as natural systems, and develop their own landscapes
- describe the global distribution of past and present cold environments, and outline their main physical characteristics
- discuss the range of geomorphological processes that operate in cold environments, including weathering, erosion, transportation and deposition by the actions of ice (glacial),

meltwater (fluvioglacial) and subaerial processes, including periglacial

- describe the variety of landscapes that have been created in cold environments over time, largely due to the actions of ice (glacial), meltwater (fluvioglacial) and subaerial processes, including periglacial
- explain the varying impacts of human activity on environmentally fragile cold environments over time, including the impact of climate change
- consider the ways in which fragile cold environments can be managed

Questions & Answers

About this section

In this section of the book, two sets of questions on each of the content areas are given, one set for AS and one for A-level. For each of these, the style of questions used in the examination papers has been replicated, with a mixture of multiple-choice questions, short-answer questions, data-stimulus questions and extended-prose questions. Other than the multiple-choice questions and some short, knowledge-based questions, all will be assessed using a 'levels of response' mark scheme to a maximum of four levels. The relative proportions and weightings of these questions varies between AS and A-level.

The sections that follow are each structured as follows:

- sample questions in the style of the examination
- mark schemes in the style of the examination
- example student answers at an upper level of performance
- examiner's commentary on each of the above

For AS and A-level Geography, all assessments will test one or more of the following assessment objectives (AOs):

- **AO1:** demonstrate knowledge and understanding of places, environments, concepts, processes, interactions and change, at a variety of scales.
- **AO2:** apply knowledge and understanding in different contexts to interpret, analyse and evaluate geographical information and issues.
- **AO3:** use a variety of relevant quantitative, qualitative and fieldwork skills to: investigate geographical questions and issues; interpret, analyse and evaluate data and evidence; construct arguments and draw conclusions.

All questions that carry a large number of marks (at AS and A-level) require students to consider connections between the subject matter and other aspects of the specification, or to develop deeper understanding, in order to access the highest marks. The former used to be referred to as synopticity, but the new term used is now **connections** — so try to think of **links** between the subject matter you are writing about and other areas of the specification. Some questions will target specific links. At **AS**, one topic is covered, worth a total of 40 marks. The breakdown of the questions per topic is:

- two multiple-choice questions (AO1 or AO3)
- one 3-mark question (AO1)
- one 6-mark question with data — marked to two levels (AO3)
- one 9-mark question requiring extended-prose responses, marked to three levels (AO1/AO2)
- one 20-mark question requiring extended-prose responses, marked to four levels (AO1/AO2)

Note: AS questions are not available for Hot desert systems and landscapes.

You should allocate *1 minute per mark* to answer the written questions.

At **A-level**, two topics are covered, worth 36 marks each. The breakdown of the questions per topic is:

- one 4-mark question (AO1)
- one 6-mark question with data — marked to two levels (AO3)
- one 6-mark question with data — marked to two levels (AO1/AO2)
- one 20-mark question requiring extended-prose responses, marked to four levels (AO1/AO2)

You should allocate *1½ minutes per mark* to answer the written questions.

For each question in this book, an answer has been provided towards the upper end of the mark range. Study the descriptions of the 'levels' given in the mark schemes carefully and understand the requirements (or 'triggers') necessary to move an answer from one level to the one above it. You should also read the commentary with the mark schemes to understand why credit has or has not been awarded. In all cases, actual marks are indicated.

The extended-prose response writing tasks at both AS and A-level, which each carry 20 marks, will be assessed using a generic mark scheme such as the one below. Study this carefully to see what is needed to move from one level to the next.

Level/mark range	Criteria/descriptor
Level 4 (16–20 marks)	■ Detailed evaluative conclusion that is rational and firmly based on knowledge and understanding, which is applied to the context of the question (AO2). ■ Detailed, coherent and relevant analysis and evaluation in the application of knowledge and understanding throughout (AO2). ■ Full evidence of links between knowledge and understanding and the application of knowledge and understanding in different contexts (AO2). ■ Detailed, highly relevant and appropriate knowledge and understanding of place(s) and environments used throughout (AO1). ■ Full and accurate knowledge and understanding of key concepts and processes throughout (AO1). ■ Detailed awareness of scale and temporal change, which is well integrated where appropriate (AO1).
Level 3 (11–15 marks)	■ Clear, evaluative conclusion that is based on knowledge and understanding, which is applied to the context of the question (AO2). ■ Generally clear, coherent and relevant analysis and evaluation in the application of knowledge and understanding (AO2). ■ Generally clear evidence of links between knowledge and understanding and the application of knowledge and understanding in different contexts (AO2). ■ Generally clear and relevant knowledge and understanding of place(s) and environments (AO1). ■ Generally clear and accurate knowledge and understanding of key concepts and processes (AO1). ■ Generally clear awareness of scale and temporal change, which is integrated where appropriate (AO1).

Level/mark range	Criteria/descriptor
Level 2 (6–10 marks)	■ Some sense of evaluative conclusion partially based upon knowledge and understanding, which is applied to the context of the question (AO2). ■ Some partially relevant analysis and evaluation in the application of knowledge and understanding (AO2). ■ Some evidence of links between knowledge and understanding and the application of knowledge and understanding in different contexts (AO2). ■ Some relevant knowledge and understanding of place(s) and environments, which is partially relevant (AO1). ■ Some knowledge and understanding of key concepts, processes and interactions, and change (AO1). ■ Some awareness of scale and temporal change, which is sometimes integrated where appropriate; there may be a few inaccuracies (AO1).
Level 1 (1–5 marks)	■ Very limited and/or unsupported evaluative conclusion that is loosely based upon knowledge and understanding, which is applied to the context of the question (AO2). ■ Very limited analysis and evaluation in the application of knowledge and understanding; lacks clarity and coherence (AO2). ■ Very limited and rarely logical evidence of links between knowledge and understanding and the application of knowledge and understanding in different contexts (AO2). ■ Very limited relevant knowledge and understanding of place(s) and environments (AO1). ■ Isolated knowledge and understanding of key concepts and processes. ■ Very limited awareness of scale and temporal change, which is rarely integrated where appropriate; there may be a number of inaccuracies (AO1).
Level 0 (0 marks)	■ Nothing worthy of credit.

Examination skills

Command words used in the examinations

Command words are the words and phrases used in exams and other assessment tasks that tell students how they should answer the question. The following command words could be used:

Analyse Break down concepts, information and/or issues to convey an understanding of them by finding connections, and causes and/or effects.

Annotate Add to a diagram, image or graphic a number of words that describe and/or explain features, rather than just identify them (which is labelling).

Assess Consider several options or arguments and weigh them up so as to come to a conclusion about their effectiveness or validity.

Comment on Make a statement that arises from a factual point made — add a view, an opinion or an interpretation. In data/stimulus-response questions, examine the stimulus material provided and then make statements about the material and its content that are relevant, appropriate and geographical, but not directly evident.

Compare Describe the similarities and differences for at least two phenomena.

Contrast Point out the differences between at least two phenomena.

Critically Often occurs before 'Assess' or 'Evaluate', inviting an examination of an issue from the point of view of a critic, with a particular focus on the strengths and weaknesses of the points of view being expressed.

Define/What is meant by State the precise meaning of an idea or concept.

Describe Give an account in words of a phenomenon that may be an entity, an event, a feature, a pattern, a distribution or a process. For example, if describing a landform, say what it looks like, give some indication of size or scale, what it is made of and where it is in relation to something else (field relationship).

Discuss Set out both sides of an argument (for and against), and come to a conclusion related to the content and emphasis of the discussion. There should be some evidence of balance, though not necessarily of equal weighting.

Distinguish between Give the meaning of two (or more) phenomena and make it clear how they are different from each other.

Examine Consider carefully and provide a detailed account of the indicated topic.

Explain/Why/Suggest reasons for Set out the causes of a phenomenon and/or the factors that influence its form/nature. This usually requires an understanding of processes.

Evaluate Consider several options, ideas or arguments and form a view based on evidence about their importance/validity/merit/utility.

Interpret Ascribe meaning to geographical information and issues.

Justify Give reasons for the validity of a view, idea or why some action should be undertaken. This might reasonably involve discussing and discounting alternative views or actions.

Outline/Summarise Provide a brief account of relevant information.

To what extent Form and express a view as to the merit or validity of a view or statement after examining the evidence available and/or different sides of an argument.

■ AS questions

Water and carbon cycles

Examples of multiple-choice questions

Question 1

Which of the following terms is **not** associated with systems? (1 mark)

   A inputs

   B feedback

   C resilience

   D flows

Question 2

The cryosphere refers to: (1 mark)

   A sea levels, clouds and sea ice

   B ice caps, ice shelves and glaciers

   C lakes, rivers and ponds

   D groundwater, springs and oases

Question 3

Which of the following is **not** part of a storm hydrograph?   (1 mark)

   A peak discharge

   B rising limb

   C lag time

   D geology

Question 4

Land use change can result in more carbon being released to the atmosphere such as in: (1 mark)

   A deforestation

   B afforestation

   C reforestation

   D sequestration

Question 5

Human-induced sequestration can include: (1 mark)

   A greater levels of tillage

   B carbon capture in power stations

   C creating more pasture from forests

   D burning fossil fuels

Answers to multiple-choice questions

Question 1

Correct answer C.   (1 mark)

Question 2

Correct answer B.   (1 mark)

Question 3

Correct answer D.   (1 mark)

Question 4

Correct answer A.   (1 mark)

Question 5

Correct answer B.   (1 mark)

Written-answer questions

Question 1

What is meant by the 'water balance'? (3 marks)

ⓔ Mark scheme: 1 mark per valid point.

> **Student answer**
>
> The water balance refers to where inputs of precipitation (P) are balanced by outputs in the form of evapotranspiration (E) and runoff (Q) together with changes to the amounts of water held in storage within the soil and groundwater (S) ⓐ. This is summarised by the formula: $P = E + Q + S$. When precipitation exceeds evapotranspiration, this produces a water surplus ⓑ. When evapotranspiration is greater than precipitation, there is a water deficit and demands are met by water being drawn to the surface of the soil by capillary action ⓒ.

ⓔ **3/3 marks awarded.** ⓐ ⓑ ⓒ The student provides three correct statements.

Question 2

Figure 1 shows actual and predicted carbon dioxide emissions per capita for a selection of countries and the world between 1965 and 2035. Interpret the information shown. (6 marks)

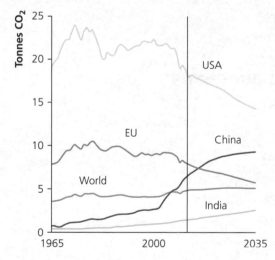

Source: BP (2014) Energy Outlook

Figure 1 Carbon dioxide emissions from energy use per capita — selected countries and world 1965–2035

(e) Mark scheme:

■ Level 2 (4–6 marks): clear interpretation of the trends shown in the figure with some qualification and/or quantification, which makes appropriate use of data to support. Interpretation may have some detail and/or sophistication.

■ Level 1 (1–3 marks): basic interpretation of the figure (likely to be largely ups and downs), with limited use of data to support.

Student answer

The striking feature of the information shown in Figure 1 is that the declines that have taken place, and will continue to take place, in the amount of carbon dioxide emissions per capita from the USA and the EU are almost totally matched by the increases in India and China. The outcome of this is that the world's total emissions per capita of carbon dioxide are only predicted to rise by a small amount, from 4 to 5 tonnes. **a**

China's per capita emissions of carbon dioxide are predicted to rise rapidly to almost 10 tonnes per capita, and this is due to the ongoing industrialisation of the country making use of cheap coal and coal-fired power stations. Similarly, in India there are signs of increase, though not as dramatic as China. This is probably due to the fact that India's development has been built on, and will continue to be built on, service industries. These are not as polluting as China's. **b**

The USA's per capita emissions are far greater than anyone else's — being at about 20 tonnes and over per capita for the last 40–50 years. This is due to its excessive use of gas-guzzling vehicles, and heavy industries pumping out lots of carbon into the atmosphere. However, it is predicted to fall to 15 tonnes per capita by 2035. This prediction may have to change due to the amount of tar shales that are being exploited for their gas and oil. This is a rapidly increasing industry at the moment. **c**

Finally, the EU's emissions of carbon dioxide per capita are also predicted to fall to about 6 tonnes. This is most likely due to these countries seeking to cut their emissions, as they are introducing more renewable sources of energy to cut down on emissions. **d**

(e) **6/6 marks awarded. a** The student begins with a sophisticated overview of the data, as well as recognising that some of the data are in the past, and some in the future. There is also some qualitative and quantitative use of the data. **b** This paragraph contains both good description and valid interpretation — evidence of geographical knowledge is being shown here. **c** Similarly there is accurate description and interpretation regarding the USA in the following paragraph. **d** The final paragraph is perhaps the weakest, but follows the same themes — the student probably felt he/she needed to refer to all elements of the figure, but this was not needed as maximum credit was already achieved.

Question 3

Evaluate how natural variations produce changes in the carbon cycle over time.　(9 marks)

ⓔ Mark scheme:

■ Level 3 (7–9 marks):

  ☐ AO1: demonstrates detailed knowledge and understanding of concepts, processes, interactions and change. These underpin the response throughout.

  ☐ AO2: applies knowledge and understanding appropriately and with detail. Detailed evidence of the drawing together of a range of geographical ideas, which is used constructively to support the response. Evaluation is detailed and well supported with appropriate evidence. A well-balanced and coherent argument is presented.

■ Level 2 (4–6 marks):

  ☐ AO1: demonstrates some appropriate knowledge and understanding of concepts, processes, interactions and change. These are mostly relevant though there may be some minor inaccuracy.

  ☐ AO2: applies some knowledge and understanding appropriately. Emerging evidence of the drawing together of a range of geographical ideas, which is used to support the response. Evaluation is clear with some support of evidence. A clear argument is presented.

■ Level 1 (1–3 marks):

  ☐ AO1: demonstrates basic/limited knowledge and understanding of concepts, processes, interactions and change. These offer limited relevance and/or there is some inaccuracy.

  ☐ AO2: applies limited knowledge and understanding appropriately. Basic evidence of drawing together of a range of geographical ideas, which is used at a basic level to support the response. Evaluation is basic with limited support of evidence. A basic argument is presented.

Student answer

I will look at how two types of natural variation produce changes in the carbon cycle over time: volcanic activity linked to plate tectonics, and wildfires.

Carbon is constantly circulating through the natural system. The carbon balance constantly fluctuates between the rates at which it leaks out (through volcanoes) and is taken in by plants (photosynthesis). Plants, through their growth, break up surface rocks such as granite, and microorganisms also hasten the weathering with enzymes and organic acids in the soil coupled with the carbonic acid and carbon dioxide in the rainwater. Over time, weathered calcium and bicarbonates are washed down to the sea and used by microscopic marine life to form shells. When organisms die these carbonate shells are deposited as carbonate-rich sediment and eventually converted to rock by pressure to form limestone. The cycle begins again when these deposits are subducted and once again brought up and exhaled by volcanoes. Whenever volcanoes erupt, such as Eyjafjallajökull in Iceland or Mount Merapi in Indonesia, amounts of carbon in the atmosphere increase greatly. ⓐ

Wildfires are common in Australia. Here, wildfires burn in areas of vegetation such as forest, bush, scrub or grassland and hence destroy huge amounts of stored carbon. Weather, together with topography and the availability of fuel, creates the unpredictable, uncontrollable and fast-moving behaviour that defines most wildfires (called 'bushfires' in Australia). However, many ecosystems and plant species in Australia evolved with fire and depend on it for germination, reproduction and habitat renewal. For example, eucalyptus, which dominates many Australian forests, has followed the spread of fire across the continent for millennia. Its growth is triggered by fire. It must also be noted that fire has been an essential land management tool for Aboriginal Australians for many centuries and is still used today. **b**

e **9/9 marks awarded.** **a** The answer refers to two natural variations that apply at two different scales. The second paragraph deals with what is known as the long-term carbon cycle and outlines the main processes involved. Knowledge and sequencing are detailed and the paragraph ends with some exemplification. **b** The third paragraph is stronger in that wildfires are clearly discussed and evaluated in the context of a specific location — Australia. Not only are the impacts on the carbon cycle clear, but there is also a sense of evaluation of how useful this natural variation is to nature and humans alike. All elements of Level 3 have been addressed in the time available — maximum marks awarded.

Question 4

Assess the relative importance of natural variation and human impact on changes in the water cycle over time.   (20 marks)

e **Mark scheme: See the generic mark scheme on pages 56–57.**

Student answer

One natural variation I have studied is a storm event, which may create flooding. Flooding is a natural occurrence that is solely dependent on the weather. The most obvious cause of flooding is a high amount of precipitation, which is greater than the rate of interception and can lead to flash floods. Also a period of precipitation over a long time can lead to a soil moisture surplus in the water budget or saturation of the ground where the ground stops absorbing water and therefore leads to overland flow and then flooding. **a** Other things such as impermeable rock in the ground or melting ice and snow can influence floods. Also, weather previous to an event of heavy precipitation can influence flooding. Very cold weather can cause the ground to become frozen and act like impermeable rock, which will make infiltration impossible, leading to a large amount of surface runoff. **b** The natural geography of areas will impact flooding, with some areas of low-lying land, such as Bangladesh, being prone to natural flooding where large rivers and tributaries meet and overflow on to floodplains. Here the monsoon season is also a major flood cause as in the monsoon season much rain falls on these large, flat drainage basins. In 2007 Bangladesh suffered some of the worst river floods in years — 169.5 mm of rain fell over 36 hours and this meant the ground was saturated and so excess water flowed straight to the River Ganges as surface runoff. **c**

Human impacts are often seen as more important in the causes of flooding. ⓓ However, flooding still took place before any human influence. A major cause of flooding is deforestation, where humans cut down trees and vegetation either for timber resources or to build upon the cleared ground. This reduces the amount of interception, which would usually lead to some evaporation. Also, trees and vegetation would absorb a large amount of soil moisture and would therefore reduce the likelihood of flooding by taking in the water and transpiring it back into the atmosphere. Humans build impermeable surfaces with roads and areas of concrete. These are essentially very similar to frozen ground in areas of permafrost such as the Alaskan North Slope or natural impermeable rock because they prevent infiltration. ⓔ

In places such as Carlisle, which lies on a floodplain, the building of roads and concrete structures resulted in severe flooding in 2005 and 2015. The build-up of impermeable surfaces such as concrete pavements means that when precipitation occurs, infiltration is not possible and rainwater flows straight to the river channel as surface runoff, decreasing lag time and increasing river discharge ⓕ — effectively causing a river to flood.

Efficient drainage systems also don't help as a human cause. Water is carried straight to the river via drains instead of gradually getting there via baseflow. An example of this was again found in Carlisle in 2005 and 2015 when the city's drainage system caused a large proportion of the flooding.

Overall, the importance of both of these elements varies according to location. It is true though that human causes can be minimalised or prevented, whereas physical variations are much more dependent on nature and difficult to control, it at all. So, in relative terms, natural variation seems to be the main factor. ⓖ

ⓔ **17/20 marks awarded.** All elements of the mark scheme for Level 4 have been addressed. ⓐⓑⓒ There is a clear account of a range of both natural variation and ⓓ human impacts on the water cycle. ⓔ The student makes a link to climate, and to cold environments — a connection to a different context. ⓕ There is also evidence of sequencing. Some references to case studies are evident too (Bangladesh and Carlisle). ⓖ Explicit assessment of relative importance is given in the conclusion. However, the answer is dominated by flooding, with a small section on land-use change (deforestation). A little more on other aspects of the water cycle, such as farming practices and water abstraction, would have achieved maximum marks. Mid Level 4 awarded.

Coastal systems and landscapes

Examples of multiple-choice questions

Question 1

Which is an output from a coastal system? (1 mark)

  A waves

  B a wave-cut platform

C tidal currents

D prevailing winds

Question 2

Tides are caused by: (1 mark)

    A the gravitational interplay between the moon and the sun

    B prevailing winds

    C the subsurface shape of a coastline

    D strong rip currents

Question 3

Which is **not** a sub-aerial process operating on a coastline: (1 mark)

    A longshore drift

    B salt weathering

    C biological weathering on a wave-cut platform

    D slumping on a cliff top

Question 4

A coastal erosional landform is: (1 mark)

    A a sand spit

    B a wave-cut notch

    C a tombolo

    D a barrier beach

Question 5

Which of the following terms is **not** associated with sea-level change: (1 mark)

    A eustatic

    B isostatic

    C tectonic

    D accretion

Answers to multiple-choice questions

Question 1

Correct answer B. (1 mark)

Question 2

Correct answer A. (1 mark)

Question 3

Correct answer A. (1 mark)

Question 4

Correct answer B. (1 mark)

Question 5

Correct answer D. (1 mark)

Written-answer questions

Question 1

Distinguish between eustatic and isostatic sea-level change. (3 marks)

ⓔ Mark scheme: 1 mark per valid point. 2 marks max. on any one element.

> **Student answer**
>
> Eustatic sea-level change is on a global level **a**, and is caused by worldwide increases or decreases in sea level as a result of ice caps melting, for example **b**. On the other hand, isostatic level change is caused by the falling or uplifting of a particular area of land in relation to the sea level due to, for example, melting glaciers. **c**

ⓔ **3/3 marks awarded. a b c** The student provides three correct statements.

Question 2

Figure 2 shows a coastal landform. Describe and explain the characteristics of the landform shown. (6 marks)

Figure 2 A coastal landform

ⓔ Mark scheme:
- Level 2 (4–6 marks): clear use of the figure to describe landform — detailed characteristics such as shape, size, geology and possible field location. Formation factors are reasonably derived from the source and are appropriate. For full marks, characteristics and formation should be integrated.
- Level 1 (1–3 marks): basic description of the feature, likely to be largely generic. Reference to formation is basic.

Student answer

The feature shown in the photograph is an arch within a cliff coastline. **a** The height of the arch is about half the height of the cliff, which is probably about 30 m tall. The arch is about 5 m wide. The upper part of the arch shows some signs of further weakening, as the rock looks a little more jointed. **b** At the base of the arch there is a clear darker area where waves will be attacking the rocks twice a day as the tide comes in. There is a distinct wave-cut notch here, too. **c** Coming back towards the photographer, there is a pebble beach with large boulders on it. These will be picked up by the powerful waves attacking the coastline and hurled at the cliff face and the walls of the arch, and eroding them with some force. **d** The area appears to be a geo, a steeply walled channel, and waves will be forced around in a swirling motion, especially if there are strong winds. This will make the erosive power, and the power of the waves themselves, much greater. **e** To the left of the photo I can see another smaller cave that will become another arch eventually.

The rocks in the photograph are clearly sedimentary as there are many layers. It is probably limestone at a place such as Flamborough Head in Yorkshire. This jointing makes it easier for the waves to exploit any weaknesses and thereby increase levels of erosion. **f**

e **6/6 marks awarded.** **a** The student provides the name of the landform and offers an accurate field relationship. **b** **c** The answer then includes detailed description of the landform, together with some integration of the processes for its formation. **d** This is followed by a more detailed account of formation and the factors affecting it. **e** Further statements linking description (field relationship) and processes follow. The account is fully integrated, as required by the mark scheme. **f** There are clear and explicit links to the figure. Maximum marks awarded.

Question 3

Assess the impact(s) of rising sea levels on coastal environments. (9 marks)

e Mark scheme:

■ Level 3 (7–9 marks):

☐ AO1: demonstrates detailed knowledge and understanding of concepts, processes, interactions and change. These underpin the response throughout.

☐ AO2: applies knowledge and understanding appropriately and with detail. Detailed evidence of the drawing together of a range of geographical ideas, which is used constructively to support the response. Assessment is detailed and well supported with appropriate evidence. A well-balanced and coherent argument is presented.

■ Level 2 (4–6 marks):
   ☐ AO1: demonstrates some appropriate knowledge and understanding of concepts, processes, interactions and change. These are mostly relevant though there may be some minor inaccuracy.
   ☐ AO2: applies some knowledge and understanding appropriately. Emerging evidence of the drawing together of a range of geographical ideas, which is used to support the response. Assessment is clear with some support of evidence. A clear argument is presented.
■ Level 1 (1–3 marks):
   ☐ AO1: demonstrates basic/limited knowledge and understanding of concepts, processes, interactions and change. These offer limited relevance and/or there is some inaccuracy.
   ☐ AO2: applies limited knowledge and understanding appropriately. Basic evidence of drawing together of a range of geographical ideas, which is used at a basic level to support the response. Assessment is basic with limited support of evidence. A basic argument is presented.

Student answer

Rising sea levels are caused by eustatic processes, which can be the melting of glaciers, which increases the seawater's volume, causing it to rise relative to the land. **a**

The impact on the coast involves a variety of landforms. First, there are submergent landforms, where the base level rises relative to the land. For example, rias, which are submerged river valleys and have gentle valley slopes and a wide river channel. An example of this is Kingsbridge Estuary in Devon, now a popular sailing point. Another submergent landform is a fjord, which is a submerged glacial valley. They have sharply steep sides and look much like a wide, deep lake. An example is the Geirangerfjord in Norway. **b**

As these happened so long ago, we have evolved with them and make use of them for our own uses, but had that happened today, the impact on our lives would be catastrophic. The British coast is already experiencing positive eustatic change, which has increased the erosion of our cliffs as the sea level is rising above the natural barrier of the beach — for example in East Yorkshire, where weaker rock such as Mappleton's boulder clay is eroding at 2 m a year, and this coastline has already lost 32 villages in the documented life of man. **c**

As coastal flooding will continue due to global warming, important coastal cities such as New Orleans, Cairo, London and Shanghai may be lost. This will impact world politics, economy and resources — for example, the Nile delta relies upon irrigation channels from the river and Egyptians could lose their cotton trade — nearly half of their income — if flooded. **d**

In conclusion, eustatic change has always occurred and as a sea-bound island, Great Britain needs to act against these negative impacts — as we used to be connected to the Netherlands and France and eventually we could end up as a collection of landforms. But it is also important to see the impact on the environment as wildlife habitats are being threatened, such as saltmarsh, home to the redshank. **e**

e **8/9 marks awarded.** **a** The answer begins with a brief introduction identifying cause. **b** The next paragraph describes the impact of rising sea levels on landforms, with some good support. **c** The following section introduces another physical impact — coastal erosion — but with a human dimension. **d** This theme is developed further into other parts of the world where the impacts are different. **e** The concluding paragraph is a little vague, though ends well with another more specific environmental impact. Overall, assessment is detailed though not always explicitly stated. Mid Level 3 awarded.

Question 4

'Soft engineering works in harmony with the natural environment and is effective in protecting the coast.' To what extent do you agree with this view?   (20 marks)

e Mark scheme: See the generic mark scheme on pages 56–57.

> **Student answer**
>
> Soft engineering is the process in which natural processes and environments are used to limit and decrease coastal erosion. The different processes involve the enhancement and modification of these coastal environments. **a**
>
> Soft engineering does involve much less erecting and building of artificial structures as is done in hard engineering, so it is working with more harmony with the natural environment. It often aids and protects natural environments. **b** An example is the great dune area in the Netherlands. Here the dune restoration and protection scheme has aided the environment and provided coastal protection. So in this case soft engineering really did cooperate with the environment. The process promoting the free actions of nature and the protection of the environment certainly did have an effect. **c**
>
> The process of beach nourishment is the process by which sand is dumped onto a beach to increase and stabilise it. This sediment will be taken from another area, so it is not in harmony with the environment. However, judging by its effectiveness, it is fairly effective from the protection aspect, as beaches are one of the best and most effective systems of defence. But, this has to be instigated often to enable it to function properly. **d**
>
> Another soft engineering technique is managed retreat, which involves the artificial or natural flooding of land. In one sense this counteracts harmony but on the other it promotes harmony and biodiversity. This managed retreat aims to establish salt marshes as low-energy environments. Examples are in Essex, in which land is surrendered and given back. This is also practised on a much greater scale in California, USA. The salt marshes that are produced often present a good nesting ground for birds. However, in terms of effectiveness is this policy seen differently? It may act as a buffer zone from coastal erosion and storms. In Essex, ponds and small lakes are found that are effective at reducing the erosional impact of the sea. **e**

The strategy of doing nothing is not in harmony with the environment, as land will be left unprotected and it will be eroded violently. Habitats can be destroyed and thereby affect natural environments. However, this strategy is also very effective. In terms of effectiveness it might be useful as coastal erosion might focus upon the designated area. This will then hopefully decrease the coastal erosion in other places. f

In my opinion, I believe soft engineering is partially working in harmony with the environment. This depends on the strategy involved. However, in its effectiveness it can either be very effective or only show low protection. g

e **17/20 marks awarded.** a The answer begins with an introduction that defines the term 'soft engineering'. b There is also clear engagement with the question at the outset. c This is followed by an example — the dunes of the Netherlands — though with little detail. d A second example is then referred to — beach nourishment. Although given in a general sense, this strategy is discussed in the context of the question. e The next paragraph considers managed retreat with some use of support, and once again the themes of the question are addressed. f Another strategy — do nothing — features in the next paragraph, again in a general sense. g The answer ends with a conclusion that summarises the points made.

Overall, a number of strategies feature in this answer, and each one is considered within the themes of the question — harmony and effectiveness. The major weaknesses of the answer, though quite coherent throughout, are the lack of detailed support and hence a restricted sense of different contexts. For the latter the student could have developed the hinted-at points regarding the importance of the habitats that soft engineering creates. Mid Level 4 awarded.

Glacial systems and landscapes

Examples of multiple-choice questions

Question 1

Which is an output from a glacial system?   (1 mark)

    A   avalanche

    B   meltwater stream

    C   snowfall

    D   firn

Question 2

Which of the following is a form of glacial movement?   (1 mark)

    A   freeze–thaw weathering

    B   regelation

    C   basal sliding

    D   nivation

Question 3

A glacial erosional landform is: (1 mark)

  A a truncated spur
  B an esker
  C a varve
  D a pingo

Question 4

Which of the following is **not** found on an outwash plain? (1 mark)

  A a sandur
  B a kettle lake
  C till
  D eskers

Question 5

A periglacial landscape typically includes: (1 mark)

  A erratics
  B a hanging valley
  C a roche moutonnée
  D stone polygons

Answers to multiple-choice questions

Question 1
Correct answer B. (1 mark)

Question 2
Correct answer C. (1 mark)

Question 3
Correct answer A. (1 mark)

Question 4
Correct answer C. 1 mark)

Question 5
Correct answer D. (1 mark)

Written-answer questions

Question 1

Explain why the snout of a glacier advances and retreats.

(3 marks)

ⓔ Mark scheme: 1 mark per valid point. 2 marks max. on any one element.

> **Student answer**
>
> The snout of a glacier advances and retreats due to its overall budget. A glacier will advance if the rate of accumulation is greater than the rate of ablation **a**, which means it is gaining more ice through snowfall than it is losing by melting **b**. A glacier will retreat if the rate of ablation is greater than the rate of accumulation. **c**

ⓔ **3/3 marks awarded. a b c** The student provides three correct statements.

Question 2

Figure 3 shows a glaciated landscape. Describe and explain the characteristics of the landscape shown.

(6 marks)

Figure 3 A glaciated landscape

ⓔ Mark scheme:

- Level 2 (4–6 marks): clear use of the figure to describe landscape — detailed characteristics such as shape, size, geology and field location. Formation factors are reasonably derived from the source and are appropriate. For full marks characteristics and formation should be integrated.
- Level 1 (1–3 marks): basic description of the landscape, likely to be largely generic. Reference to formation is basic.

Student answer

There are corries present on the far upper slopes of the area shown, as this is where it is colder, and snow and ice will have gathered here. **a** There is a visible knife-edged arête on the right of the picture and in the background. **b** Glaciers or snowfields mainly occupy the upper slopes as well and there is a glacier coming down from the summit on the left-hand side of the photo. **c** There is also a ribbon lake visible in the foreground, which is several kilometres in length and around 1 km wide. **d**

The landscape was formed when a powerful glacier eroded through the main valley, which is steep sided and wide bottomed. The glacier carried lots of debris and therefore had a powerful erosive force through abrasion. This eroded through the interlocking spurs that the previous river would have wound around, forming a misfit stream and truncated spurs. **e**

Extra erosion occurred where there was compressing flow due to increased pressure forming a deep rock basin. This is done by abrasion that occurs at the bottom of the glacier, where rocks embedded within the glacier scrape against the valley floor and sides, having a sandpapering effect. **f**

Abrasion is more effective in warm-based glaciers, e.g. the Alps, which this photo appears to show. The hanging valleys on the right-hand side were formed due to differential erosion where the tributary glaciers didn't have as much power to erode down the main glacier. **g**

e **6/6 marks awarded. a b c d** The student provides the name of a number of landforms within the landscape and offers an accurate description together with some field relationships. **e** In the second paragraph, the student provides some description of the glacial trough, together with some integration of the processes for its formation. **f** This is followed by a more detailed account of formation and the factors affecting it. The account has elements of integration as required by the mark scheme. **g** There are also clear and explicit links to the figure. Maximum marks awarded.

Question 3

Assess the impact(s) of economic activity on cold environments. (9 marks)

e Mark scheme:

■ Level 3 (7–9 marks):

　□ AO1: demonstrates detailed knowledge and understanding of concepts, processes, interactions and change. These underpin the response throughout.

　□ AO2: applies knowledge and understanding appropriately and with detail. Detailed evidence of the drawing together of a range of geographical ideas, which is used constructively to support the response. Assessment is detailed and well supported with appropriate evidence. A well-balanced and coherent argument is presented.

■ **Level 2 (4–6 marks):**
  ☐ **AO1:** demonstrates some appropriate knowledge and understanding of concepts, processes, interactions and change. These are mostly relevant though there may be some minor inaccuracy.
  ☐ **AO2:** applies some knowledge and understanding appropriately. Emerging evidence of the drawing together of a range of geographical ideas, which is used to support the response. Assessment is clear with some support of evidence. A clear argument is presented.
■ **Level 1 (1–3 marks):**
  ☐ **AO1:** demonstrates basic/limited knowledge and understanding of concepts, processes, interactions and change. These offer limited relevance and/or there is some inaccuracy.
  ☐ **AO2:** applies limited knowledge and understanding appropriately. Basic evidence of drawing together of a range of geographical ideas, which is used at a basic level to support the response. Assessment is basic with limited support of evidence. A basic argument is presented.

Student answer

In the Southern Ocean, whaling and fishing have taken place since the 1800s. However, there was a big problem, as by 1980 there were very few seals and whales left, since they had been farmed for their oil and bones. However, by 1998 this was regulated and now whaling can only be done for scientific research, which many countries say they do, like Norway and Japan. They are only allowed to catch around 100 whales a year as, due to climate change, the replacement rate of whales is low via breeding **a**, making this a much more sustainable activity than before as 5,000 seals and whales were dying each year; now only around 75 die in 1 year **b**.

In Alaska, however, oil exploration has become a big part of the local economy, making up nearly 75% of the economy. However, many strategies have been put in place along the pipeline to achieve a sustainable environment, as the pipeline has to travel around 1,200 km from Prudhoe Bay. **c** The main strategy is that the pipeline is raised above the permafrost to prevent the active layer from melting and allowing the caribou herd to migrate underneath the pipeline. **d** In places where this isn't possible, like mountain ranges, it is encased in an insulated box that prevents the oil freezing and the permafrost melting.

Another strategy is because the pipeline crosses fault lines, meaning it is prone to earthquakes, it is placed on a zigzag rail that moves the pipeline with the earth and prevents it from breaking and leaking oil, and damaging the environment. **e** Tourism in the north of Alaska is also more sustainable as there are buses that transport tourists to see the sites, reducing any extra wear on the ground or additional carbon dioxide emissions. **f**

In conclusion, these areas have become much more sustainable than they were before as people can now see that these environments need protecting in the long term and not just short term, and that these lands belong to everyone, but it is also a way of life for some.

(e) **8/9 marks awarded.** The student has focused on the impacts of economic activity, as required by the question — whaling in the Southern Ocean and oil extraction in the Arctic. (a)(b)(c)(d) Support is provided for both, and there are several areas of assessment. (e)(f) The account is not as well structured as it could be, as there are some weaker elements. The student has also concentrated on the theme of sustainability, which is a reasonable thing to do. Mid Level 3 awarded.

Question 4

Evaluate the varying ways in which ice moves.   (20 marks)

(e) **Mark scheme:** See the generic mark scheme on pages 56–57.

> **Student answer**
>
> One way in which ice moves is by basal sliding. Basal sliding occurs in warm, temperate glaciers. It occurs when meltwater forms at the bottom of the glacier, acting as a lubrication agent for the movement of the ice. The meltwater comes from the glacial streams, which are supraglacial and englacial and can reach the bottom of the glacier (subglacial) if they flow downwards through crevasses. It can also occur deep in the ice, as pressure melting at below 0°C causes ice to turn to water by compression. If compression is high at 2,000 m deep into the glacier, ice can melt at −1.6°C. The meltwater then allows the glacier to slide across the underlying rock with less friction. Basal sliding occurs more quickly if the slope is steeper, so there is more gravity and downwards force. (a) Basal sliding can form the major component of glacier movement in temperate glaciers as there are plentiful amounts of meltwater to facilitate it. (b)
>
> Another way ice moves is by internal flow, where the sheer pressure from ice above compacting the ice down allows linear ice crystals to be formed. These ice crystals are aligned in the direction of glacial movement. This allows them to slide over one another using intergranular slip, which allows the ice to move. The rigid zone above the crystals does not move but is carried along as if being 'piggy-backed' by the ice below. (c) This form of movement is the main way in which ice in polar glaciers move as there is little, if any, meltwater. (d)
>
> In addition, ice can also move by compressional and extensional flow. In extensional flow the ice spreads downslope and forms crevasses as it spreads out down a steep slope, and in compressional flow the ice slows and compresses and compacts together, tightening the crevasses and forming bulges on the surface of the glacier. (e) This type of movement can occur in both polar and temperate glaciers, especially where there is a steep gradient to flow down. It could occur in a narrow valley glacier and it could also occur at the edge of a polar glacier, such as when there is undercutting by seawater at the edge of an ice shelf in the Antarctic peninsula. (f)
>
> Rotational flow occurs in corries and is when the ice in the hollow moves round an imaginary central pivot point from the back wall to the rock lip of a corrie. (g)

ⓔ 18/20 marks awarded. This answer is technically accurate and focused on the question. ⓐⓒⓔⓖ Four different methods of ice movement are featured, the first three being more detailed than the last. In all cases the processes are explained fully with detailed sequencing allowing the answer to access Level 4. ⓑⓓⓕ Furthermore, each of the first three paragraphs ends with an evaluation of the importance of that method of ice flow for particular and different contexts. The lack of detail for the last method prevents the answer from gaining full marks — perhaps the student was running out of time and had to rush. Mid Level 4 awarded.

■ A-level questions

Water and carbon cycles

Written-answer questions

Question 1

What is meant by the carbon budget? (4 marks)

ⓔ **Mark scheme: 1 mark per valid point.**

> **Student answer**
>
> The carbon budget refers to the net balance of carbon exchanges between the four major stores of carbon (the lithosphere, hydrosphere, atmosphere and biosphere). **a** Scientists state that there is a net gain of 4.3 Gt of carbon per year into the atmosphere **b**, mainly through rising levels of carbon dioxide and other greenhouse gases such as methane **c**. This is primarily because of emissions from the combustion of fossil fuels and cement production. **d** The land and oceans together act as sinks, with a net storage of approximately 2.5 Gt per year each, but this is not counterbalancing the gains from fossil fuel burning. **e**

ⓔ **4/4 marks awarded. a b c d e** The student provides several correct statements.

Question 2

Table 1 shows the distribution of the world's water by source. Interpret the information shown. (6 marks)

Table 1 Distribution of the world's water, by source

Water source	% of freshwater	% of total water
Oceans and seas	—	96.5
Ice caps, glaciers and permanent snow	68.7	1.74
Groundwater — freshwater	30.1	0.76
Groundwater — saline	—	0.93
Soil moisture	0.05	0.001
Ground ice and permafrost	0.86	0.022
Lakes — freshwater	0.26	0.007
Lakes — saline water	—	0.006
Atmosphere	0.04	0.001
Swamp water	0.03	0.0008
Rivers	0.006	0.0002
Biological water	0.003	0.0001

Source: United States Geological Survey (USGS)

ⓔ Mark scheme:

■ Level 2 (4–6 marks): clear interpretation of the data in the table with some qualification and/or quantification, which makes appropriate use of data to support. Interpretation may have some detail and/or sophistication.

■ Level 1 (1–3 marks): Basic interpretation of the data in the table (likely to be largely highs and lows), with limited use of data to support.

Student answer

The first thing to note is that the vast majority of water on this planet is stored within the oceans, over 96% in fact. This water is not potable and is saline in nature. ⓐ Another 1% (or nearly) consists of saline groundwater, which again cannot be used for drinking purposes. ⓑ This leaves only a relatively small amount that can be used for drinking — about 2.5%.

The majority of this freshwater is tied up in the cryosphere, as ice caps, glaciers and permanent snow — over two-thirds of the freshwater. Once again much of this is unavailable for people to use unless it is melted or melts naturally. Actually, this is unlikely, as the great ice caps of Antarctica and Greenland are unlikely to ever melt. ⓒ

So the freshwater that people can use — from rivers, groundwater and as precipitation from the atmospheric store — is a very small in amount indeed. Any small change in these stores, possibly from climate change or from human pollution activities, could have dire consequences. Life on Earth, whether human or animal, greatly depends therefore on these stores and the changes that may take place within them. ⓓ

ⓔ **6/6 marks awarded.** The student begins by referring to total water before moving on to freshwater. In this way he/she demonstrates full understanding of the data provided in Table 1. ⓐⓑⓒ Good qualitative statements are made and each is supported by valid comment. Appropriate terminology is also used throughout to demonstrate clear understanding and application. ⓓ Sophistication is also demonstrated by linkage to consequences of change in the water stores. Maximum marks awarded.

Question 3

Brisbane has a long history of flooding, dating back to the 1840s, when records began. Until 2011 the most devastating flood had occurred in January 1974. The Wivenhoe dam was built in the early 1980s in response to the devastation caused by the 1974 flood. In 2011 the rainfall was more prolonged and of greater intensity. It is estimated that in the 7 days leading up to the 2011 flood, the Brisbane catchment received 40% more rainfall than during the equivalent period in 1974. All this rain meant that management of water releases from the Wivenhoe dam became a critical issue. Some flood engineers believe that earlier water releases from the dam were too small, so later releases were much greater than should have been required. A massive release on 11 January was in large part responsible for flooding in Brisbane.

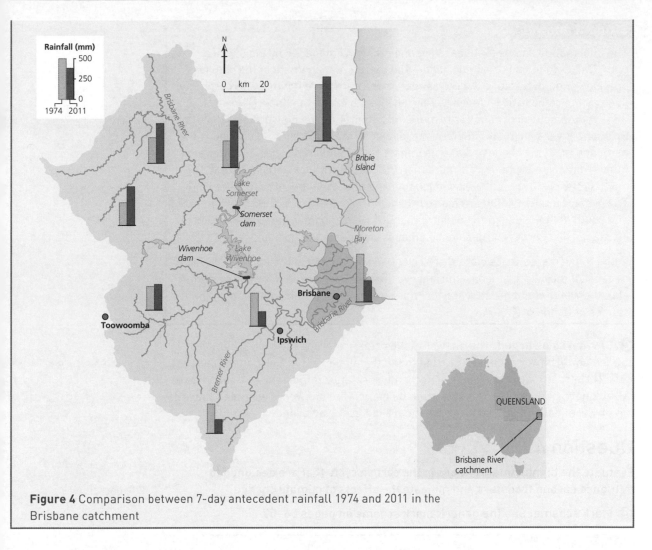

Figure 4 Comparison between 7-day antecedent rainfall 1974 and 2011 in the Brisbane catchment

The information above provides information about flooding in Brisbane, Australia in January 2011. Using this and your own knowledge, assess the relative importance of physical and human causes of this flooding event. (6 marks)

ⓔ Mark scheme:

■ Level 2 (4–6 marks):

  ☐ AO1: Demonstrates clear knowledge and understanding of concepts, processes, interactions and change.

  ☐ AO2: Applies knowledge and understanding to the novel situation, offering clear analysis and assessment drawn appropriately from the context provided. Clear statement of relative importance.

■ Level 1 (1–3 marks):

  ☐ AO1: Demonstrates basic knowledge and understanding of concepts, processes, interactions and change.

  ☐ AO2: Applies limited knowledge and understanding to the novel situation, offering basic analysis and assessment drawn from the context provided. Tentative or no statement of relative importance.

The information shows that Lake Wivenhoe is highly managed by the use of a dam. One human cause is important — the overall management of the dam was not good enough because the releases from the dam were not managed properly, such as the release of 11 January, which helped to cause the flood. **a** However, I think that the physical causes are more important **c**, as the figure shows that the Brisbane River has many tributaries leading to it, many of which do not link to the dam, and these will have had a large impact on the flooding, especially after a major rainstorm event — 40% more than the previous major flood in 1974. **b** The north of the catchment also had much more rainfall than that previous event. **b** The greater intensity of the rain also meant that much of the ground was saturated and could not take any more water, with much overland flow. This will have led to increased discharge, increasing the amount of flooding that took place. **b**

In the flooded area there were many rivers and this added to the cause of flooding because if they were all bank-full and overflowed there would have been nowhere else the water could go, showing that this physical aspect was important in the causes of flooding. **c**

e **5/6 marks awarded.** The answer makes clear reference to both **a** human causes and **b** physical causes applied to the flooding event in Brisbane in January 2011. **c** There are also two clear statements of relative importance. All aspects of Level 2 have been met — the only issue being that the answer is weighted towards the physical causes and so lacks balance. Mid Level 2 awarded.

Question 4

Evaluate the human interventions in the carbon cycle that are designed to influence carbon transfers and mitigate the effects of climate change.   (20 marks)

e Mark scheme: See the generic mark scheme on pages 56–57.

There have been several human interventions to reduce the increasing levels of carbon dioxide in the atmosphere and the resultant climate change. One of the major and most significant interventions to mitigate climate change was the Kyoto Protocol in 1997, where countries agreed that they must reduce their carbon emissions, with 1990 emissions as a baseline. **a** The Kyoto Protocol came into effect in 2005 and by 2006 it had been ratified by 162 countries. However, the USA refused to sign the treaty because President Bush believed it would be harmful to the US economy. Another flaw in the treaty was that when it had been signed in 1997, China wasn't seen as a global industrialised nation and therefore wasn't required to reduce its carbon emissions even though now it is at the forefront of greenhouse gas output. **b** The Kyoto Protocol agreed a system of carbon credits with countries being given a set limit on the amount of greenhouse gases they could release into the atmosphere. Countries that did not meet their quotas could sell their credits to other countries. However, if a country still exceeded its quota, action could be taken against it, including fines, making it invest in green technology, which could involve other countries.

In 2000, the Japanese government paid the New South Wales government $500 million, which was used to plant 40,000 hectares of forest over the next 20 years. Even though this system seems to be effective it has come under heavy criticism because it is biased towards developed countries that can just pay fines easily and also limits the growth of developing countries by stopping industrialisation. **c** In 2015 the Kyoto Protocol expired and new ways are being sought to reduce global emissions.

Then in 2007 talks for the next agreement began and 2 years of talks led up to the Copenhagen Conference at the end of 2009. Here, countries agreed that they could no longer stop global warming but they could minimise its scope and its effects. One such way was the provision of a financial package for developing countries worth $100 billion by 2020, which could be invested in green technology in these countries. Another summit took place in 2015 in Paris. **d**

At a much smaller scale, there have been several initiatives in cities around the world to introduce less polluting forms of transport and these have been successful to a limited extent. In London, for example, there are now many hybrid buses combining diesel with electricity to reduce carbon emissions, and also the well-known Boris bikes financed by Santander Bank. These are also common in many other cities around the world, such as Barcelona. **e**

To conclude, there have been many responses in an attempt to combat global warming. However, the most successful has clearly been the Kyoto Protocol, with other summits not being able to form concrete deals on paths towards reducing our global emissions.

e **19/20 marks awarded.** **a** The student discusses human interventions in the carbon cycle at two different scales: global and city-based. The answer is dominated by the global (the Kyoto Protocol), but this is to be expected. **b** For this intervention, clear detail is provided with several aspects of evaluation, such as the countries that did not sign it and the carbon credit system, with some good support regarding New South Wales and Japan. **c** Also within this paragraph the student makes links to economic development. **d** The second paragraph brings this form of intervention up to date, and in doing so demonstrates chronology. **e** The third paragraph on city strategies is more descriptive, but is valid, and contains a further brief evaluation. High Level 4 awarded.

Hot desert systems and landscapes
Written-answer questions

Question 1

Describe the climate of hot desert environments. (4 marks)

ⓔ Mark scheme: 1 mark per valid point. 2 marks max. for any one element.

Student answer

High diurnal temperatures ranging up to 30°C. ⓐ Can reach temperatures as high as 50°C in the shade in the daytime. ⓑ Very little rain — less than 250 mm per year. ⓒ When it does rain it is usually hard and intense. No clouds in the sky mean the sun is intense. ⓓ

ⓔ **4/4 marks awarded.** ⓐ ⓑ ⓒ ⓓ Although the answer is extremely concise, it provides four correct statements.

Question 2

Figure 5 shows the changes in the extent of Lake Chad, Africa, between 1963 and 2001. Interpret the changes shown in Figure 5. (6 marks)

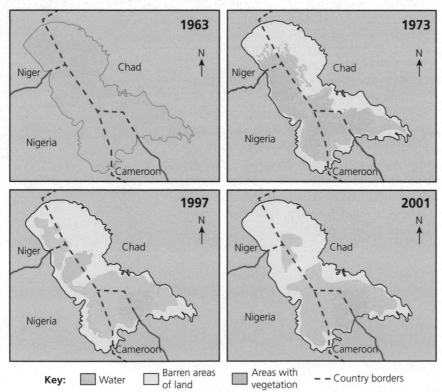

Source: UNCCD

Figure 5 The changes in the extent of Lake Chad, Africa

@ Mark scheme:

■ Level 2 (4–6 marks): clear interpretation of the changes shown in the figure with some qualification and/or quantification, which makes appropriate use of data to support. Interpretation may have some detail and/or sophistication.

■ Level 1 (1–3 marks): basic interpretation of the figure (likely to be largely increases and decreases), with limited use of data to support.

Student answer

Figure 5 shows that Lake Chad has reduced significantly in size by 2001, almost to a tenth of the size it was in 1963. a This has been accompanied by an increase in the desert around it, the Sahara spreading across where the water used to be, creating a large area of barren land. There are several areas of vegetation, usually occupying the land where the lake used to be — the ground is probably still moist here, or more likely the water table is close to the surface. b Vegetated areas now cover about a third of the area of the former lake. c

Nigeria and Niger used to have extensive parts of Lake Chad, but now they have lost all of what they had. d The fishermen that would have depended on the lake for their livelihoods, and their food, will have had to leave the area or face starvation. e Lake Chad now only exists in Cameroon and Chad — where its name originates.

@ **5/6 marks awarded.** a c d The student has provided detailed evidence of the changes shown, with clear interpretation. b e The answer also includes commentary on two aspects of the changes. To gain maximum marks the student could have either provided a little more detail of changes within the time periods, or commented on the processes causing the changes. Mid Level 2 awarded.

Question 3

Namibia is the most arid country south of the Sahara, receiving an average of only 258 mm of rain a year. With a population density of only 2 people per km, it is one of the least populated countries in the world. Namibia is divided into three topographical regions: the western coastal zone with its intensely dry Namib Desert; the eastern desert zone; and the semi-arid areas of the central plateau, where desertification is a barrier to economic progress.

Drought dominates the climate, although there is some rain during the summer. Daytime temperatures in summer can reach up to 40°C, but the Benguela ocean current cools the coast. With over 300 days of sunshine a year, tourism potential is great. Tourism is seasonal, with peaks corresponding to holidays in Europe and South Africa, when temperatures are lower and flash flooding is less of a risk.

Tourism makes up a greater proportion of the Namibian economy than it does in sub-Saharan Africa or in the world as a whole. The number of international arrivals has increased steadily since Namibian independence in 1990, despite falling at times of global economic crisis. The main attractions for tourists are Namibia's natural environment, diverse cultures and archaeological sites. About 40% of the wildlife is in protected areas, but much of it is on private commercial farms. A multi-million pound tourism industry has been established on the privately owned semi-arid to arid rangelands. It is based on safaris to view game animals, and their controlled hunting, known as trophy hunting.

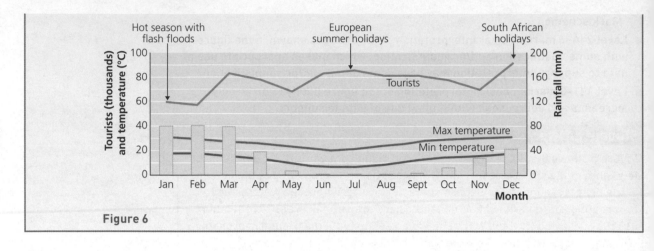

Figure 6

The above information is about tourism in the desert country of Namibia. Using this information and your own knowledge, assess the role of, and potential for, tourism for Namibia

(6 marks)

(e) Mark scheme:

■ Level 2 (4–6 marks):

☐ AO1: Demonstrates clear knowledge and understanding of concepts, processes, interactions and change.

☐ AO2: Applies knowledge and understanding to the novel situation, offering clear analysis and assessment drawn appropriately from the context provided. Connections and relationships between different aspects of study are evident, with clear relevance. Clear statement of role and/or potential.

■ Level 1 (1–3 marks):

☐ AO1: Demonstrates basic knowledge and understanding of concepts, processes, interactions and change.

☐ AO2: Applies limited knowledge and understanding to the novel situation, offering basic analysis and assessment drawn from the context provided. Connections and relationships between different aspects of study are basic, with limited relevance. Tentative or no statement of role and/or potential.

Student answer

Namibia is a country that can really harness its potential for tourism. a It has a hot and dry climate, especially during their winter (which coincides with the European summer) and there is a lot of safari-type holidays available, with the large amount of game that inhabit the sparsely populated arid landscapes. The tourism industry makes up a greater proportion of the GDP than every other sub-Saharan country, and so it has an important role to play. b The trouble is, though, that this over-dependency on tourism could work against it should the tourist trade stop or reduce for any reason. For example, the tourist industries of Egypt and Tunisia have suffered greatly in recent years due to terrorist acts. c

Namibia is also in a good location for tourists from nearby South Africa during the summer in this hemisphere as well as the northern hemisphere summer. From the graph, although it can be seen that there is a consistent number of tourists all year round, there do appear to be small peaks of over 80 000 visitors during these times. The lowest amount of tourists coincides with the period of hot weather with storms, which may cause flash floods. This is the New Year for people around the world. **d**

So it is fair to say that Namibia does rely on tourism heavily **b**, but it has the potential to exploit it even further with its climate and natural resources **a**.

e **6/6 marks awarded.** **a** The answer states at the outset that there is potential for tourist development within Namibia, which is then repeated in the brief conclusion. **b** Following a clear summary of the advantages that Namibia possesses for tourism, the student then makes a clear statement of the role of tourism in the country, again repeated in the conclusion. **c** Evidence of connections to other aspects of the subject is provided at the end of the first paragraph, with a sophisticated comment regarding over-dependence. **d** There is clear understanding of the data throughout, particularly in the second paragraph. Maximum marks within Level 2 are awarded.

Question 4

Assess the role of water in the formation of a desert landscape. (20 marks)

e Mark scheme: See the generic mark scheme on pages 56–57.

Student answer

Inselbergs are relic outcrops of rocks, which look like rounded hills but with steep slopes. They are formed when the original plateau is eroded by water. In previous times when water was more abundant, rivers will have flowed through the area, eroding the plateaux by abrasion and causing lateral and vertical erosion. These river channels are widened and deepened until eventually only resistant hard rock remains standing. These residual resistant areas of rock form inselbergs. **a**

Salt lakes form when an endoreic river terminates at a lowland area in a desert. The water cannot reach the sea so remains in the form of a lake. Due to the high temperatures in the desert, the water evaporates, leaving behind any dissolved salts. Capillary action also brings salt to the surface. This forms a lake with a high salt content. Salt lakes are lakes that may become dry with their beds covered with a salt crust in times of high evaporation — called salt flats. The salt crusts may also crack in a polygonal pattern. **b**

Wadis are steep-sided river channels in which ephemeral streams will temporarily flow. They form after a period of intense rainfall. The precipitation may collect and form channels. Due to the high volume of discharge the river has lots of erosive power. The riverbed and banks are eroded by water abrasion. This vertical and lateral erosion of the river channel creates a channel with steep sides and a flat bottom. When the water evaporates or infiltrates, the sediment that the river carries is dropped, leading to a braided channel and a dry riverbed covered with sediments. **c**

An alluvial fan is a fan-shaped area below the exit or mouth of a wadi that is composed of alluvial deposits. The largest particles are found at the mouth of the wadi whereas the smallest particles travel to the edge of the fan. They form when an ephemeral river exists in a wadi and flows onto a gently sloping area of land. The ephemeral stream transports a large volume of sediment by suspension and saltation. When the stream flows onto the gently sloping land it loses velocity and no longer has the energy to transport its load. The sediment is deposited, causing the channel to braid and distributaries to form. These distributaries spread out the alluvial deposits into a fan shape. **d**

All of these features of a desert landscape — inselbergs, wadis, alluvial fans and salt flats — have all been formed by water in an environment where wind is also important. All that wind does for these landforms is to modify what water has created, so for these landforms water is the dominant process. **e**

e **20/20 marks awarded.** **a** **b** **c** The student provides three good paragraphs at the outset that interconnect process and landform. Assessment of the role of water for each of these landforms is clear. **d** The fourth paragraph goes a little beyond this by linking the landforms of wadi and alluvial fans into a landscape, demonstrating their wider field relationship and assessing how water affects both. **e** The final paragraph provides the clear assessment element of the question. Maximum credit awarded.

Coastal systems and landscapes
Written-answer questions
Question 1

Explain how soft engineering could protect a coastline.   (4 marks)

ⓔ Mark scheme: 1 mark per valid point. 3 marks max. for any one strategy.

> **Student answer**
>
> Soft engineering can be used in a number of ways to protect a coastline. One method employed recently at Medmerry in Sussex involves the removal of old hard engineering strategies such as the sea walls. **a** This allows water to enter into the low-lying areas of the coast, creating a tidal wetland area **b**, which absorbs wave energy and stops the erosion of the land behind **c**.
>
> Another method of doing this is dune replenishment. As dunes naturally protect the coast from erosion by extending the beach and raising it, the use of vegetation to encourage dune development is a useful way of protecting the coast from high tide erosion and coastal flooding. **d**

ⓔ 4/4 marks awarded. a b c d The student provides four correct statements.

Question 2

Figure 7 shows the coastal environment around Gazi Bay, Kenya. Analyse the risks and opportunities for people in the area shown.   (6 marks)

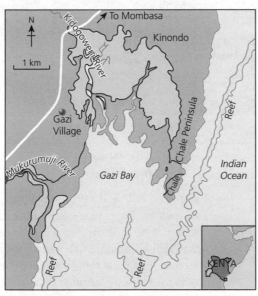

Key: ▨ Mudbanks ▨ Mangroves ☐ Coral reefs

Source: UNEP

Figure 7 Gazi Bay, Kenya, Africa

ⓔ Mark scheme:

■ Level 2 (4–6 marks): clear interpretation of the features shown in the figure with some qualification and/or quantification, which makes appropriate use of data to support. Interpretation may have some detail and/or sophistication.

■ Level 1 (1–3 marks): basic interpretation of the features shown in the figure, with limited use of data to support.

Student answer

Gazi Bay provides a number of risks for the people of the area. First, the area lies on the east coast of Kenya and as sea level rises with global warming, the land is likely to be inundated by rising levels of water. This will cause people to move out and rebuild their village and their lives. There are mudbanks near to where the village is located, which suggests that the land is flat and therefore the sea will easily cover the land and the village. a

Another risk the area faces by being on the edge of the Indian Ocean is tsunamis. When the 2004 south Asian tsunami took place there were some areas on the east coast of Kenya affected, though obviously not as severely as in Banda Aceh. The people will therefore have to have tsunami shelters, consisting of raised platforms above the rest of the land. b

Gazi Bay also has a number of opportunities as shown by the map. There are coral reefs off the coast that will encourage tourist activity and bring welcome revenue to the people of this developing country. c

Also, with the mangroves being extensive to the north of the bay, there could be some interesting flora and fauna, again encouraging a certain type of ecotourist activity to see and possibly live within this distinctive natural environment. d The people of Gazi will have to make sure they manage this activity well.

ⓔ **6/6 marks awarded.** The student has responded to all elements of the question. a b There are two risks described and analysed, and c d two opportunities described and analysed. This is a well-structured answer. Full marks awarded.

Question 3

The pebble ridge at Westward Ho! extends northwards across the mouth of the Taw estuary and is an excellent example of a rare, narrow, pebble spit. It has retreated steadily over the last 100 years and is now much more modified by artificial gabions (wire baskets filled with large stones). Although its origins are unknown there appear to be three main natural sources for the pebbles:

■ the raised beach to the west of Westward Ho! from which pebbles are still
  eroded and moved by longshore drift to the ridge
■ offshore pebbles deposited by the waves

■ materials delivered directly from erosion of the cliffs to the west

Between 1932 and 1991 the pebble ridge retreated 42 metres. The concern is that the natural supply of material from further west is not enough to sustain the ridge without human intervention. This takes the form of an annual movement of pebbles and cobbles onto the ridge by Torridge District Council. If this were to stop, it is likely the ridge would retreat by 1–2 metres per year and be breached at regular intervals. This would inundate the area of the Northam Burrows Country Park, as well as the Skern saltmarsh.

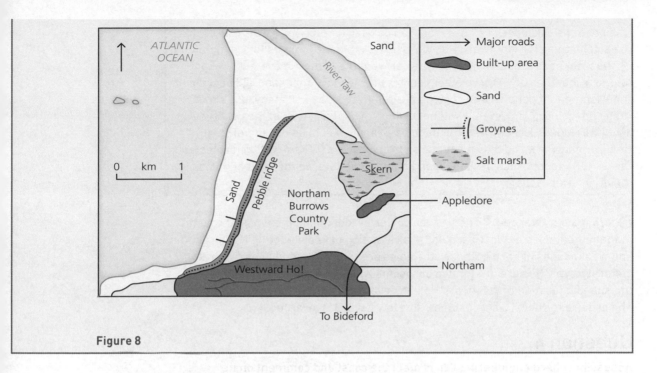

Figure 8

Using the above information and your own knowledge, assess the strengths and weaknesses of the coastal management strategies employed at Westward Ho!   (6 marks)

e Mark scheme:

■ Level 2 (4–6 marks):
  ☐ AO1: Demonstrates clear knowledge and understanding of concepts, processes, interactions and change.
  ☐ AO2: Applies knowledge and understanding to the novel situation, offering clear analysis and assessment drawn appropriately from the context provided. Clear statements of strengths and/or weaknesses.

■ Level 1 (1–3 marks):
  ☐ AO1: Demonstrates basic knowledge and understanding of concepts, processes, interactions and change.
  ☐ AO2: Applies limited knowledge and understanding to the novel situation, offering basic analysis and assessment drawn from the context provided. Tentative or no statements of strengths and/or weaknesses.

Student answer

From the text and diagram, it is clear that three forms of coastal management have been used at Westward Ho! These are gabions, groynes and a form of beach nourishment where pebbles and cobbles have been artificially placed on the natural ridge to replace the ones being eroded or moved by waves. **a**

Gabions (large stones contained within wire baskets) are designed to break up the oncoming waves and reduce their erosional power. They can be quite effective, but to many they are ugly and can be dangerous for people clambering over them. **b** Wooden groynes trap sediment moving along a beach by longshore drift and maintain the existence of a beach that can be used for tourism. [b] This is likely to be a good thing for Westward Ho!, which is trying to maintain its tourist potential. [c] The weakness though is that groynes prevent the natural movement of sand along a coastline and starve other resorts downwind of that sand. [b] The beach nourishment of pebbles and cobbles is a relatively cheap strategy, but it has to be on-going. [b] However, it will not change the natural look of the coastline [b] — the ridge is already made of pebbles — and it has to be done to protect the Country Park and Skern saltmarsh from erosion. [c] These will be protected areas for aspects such as wildlife and natural habitats, so the protection has to be done effectively. [b]

[e] **4/6 marks awarded.** [a] The answer begins by identifying the types of coastal management used at Westward Ho! — a clear sign of knowledge. [b] The strengths and weaknesses of each of these strategies are then examined in turn, in rather general terms. [c] There are limited occasions when the student applies these to the specific location of Westward Ho!, and even here they are weak statements. The criteria for Level 2 are just met, so a low Level 2 is awarded.

Question 4

Assess how hard engineering can protect the coast and comment on its effectiveness.

(20 marks)

[e] **Mark scheme: See the generic mark scheme on pages 56–57.**

Student answer

Hard engineering can be used to protect a coastline from coastal flooding and coastal erosion. One area in the UK where hard engineering is used is along the Holderness coast where cliffs are currently retreating at 2 m per year. Erosion is very prominent here because the cliffs are generally made of weak clays or till, materials that can be easily eroded, and also the sometimes large northern fetch of the waves makes them very destructive, removing material from the coast and depositing it at Spurn Head, a spit further down the coast. [a]

In Mappleton, on the Holderness coast, hard engineering protects the coast. In 1991, two rock groynes were put in place. These stop eroded sediment from moving down the coast and mean that this trapped sediment can be used to absorb wave energy, which would normally erode the cliff. Also at Mappleton, a 500 m rock revetment has been put in front of the cliff. This stops the waves hitting the cliffs and prevents hydraulic action and abrasion from eroding them. However, further down the coast, past the revetment, erosion has been increased. Therefore this hard engineering is very effective in the area it protects but can make coastal erosion worse further down the coast. [b]

Another area on the Holderness coast that uses hard engineering is around the Easington Gas Terminal, which is surrounded by a sea wall. This sea wall is successful in protecting the gas terminal because the wall reflects the energy from the waves. However, this sea wall will only protect the coast for 50/60 years. After this it may need to be replaced, but on the other hand, as supplies of gas from the North Sea decline, that may not be the case. Hard engineering is a very expensive process and it may well be that what is important today may not be so in the future (especially if the use of fossil fuels declines). Establishing priorities is important in coast management, hard or soft. c

Hard engineering is very successful and effective at protecting the coast in Holderness. However, areas that are not protected often have increased erosion and the defences must be replaced at least every 50 years. d

ⓔ **19/20 marks awarded.** a The student sets out the rationale for hard engineering in this introduction. b This is followed by good use of case study material referring to two examples of hard engineering at one location with some comment on effectiveness. c The third paragraph again makes good use of support material, with clear assessment. This paragraph also demonstrates deeper thinking with a clear set of links to other parts of the specification — climate change and energy resources. d The answer ends with a clear conclusion with an explicit sense of assessment. High Level 4 awarded.

Glacial systems and landscapes

Written answer questions

Question 1

Explain the formation of different types of moraine. (4 marks)

ⓔ Mark scheme: 1 mark per valid point. 3 marks max. on any one type.

> **Student answer**
>
> Lateral moraine is found at the side of a glacier, alongside the valley. It got there from the valley sides above the glacier. Freeze–thaw action, where water gets into cracks during the day and freezes and expands at night, breaks off bits of rock that fall down onto the glacier. **a** If two glaciers flow together then the lateral moraines come together in the middle of the main glacier to form a medial moraine. **b**
>
> Some of the rock on the surface of the glacier falls down crevasses and becomes englacial moraine. If it reaches the floor of the valley it is basal moraine. **c** The glacier carries the moraine down to the snout, where the ice is melting. If the glacier stays in the same place for a long time a ridge builds up. If it is at the furthest point the ice reached then it is a terminal moraine. **d**

ⓔ **4/4 marks awarded.** **a b c d** The student provides at least four correct statements.

Question 2

Figure 9 shows the distribution of cold environments. Analyse the distribution shown. (6 marks)

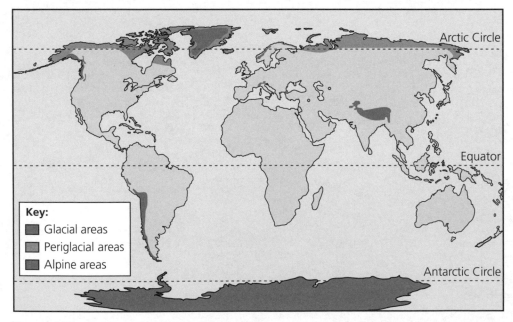

Figure 9 The distribution of cold environments

ⓔ Mark scheme:

- Level 2 (4–6 marks): clear analysis of the distribution shown in the figure with some qualification and/or quantification, which makes appropriate use of data to support. Analysis may have some detail and/or sophistication.
- Level 1 (1–3 marks): basic analysis of the figure (likely to be names of areas), with limited use of data to support.

Student answer

Cold environments include the alpine areas, e.g. the Alps and the southwest coast of South America, in parts of Chile. Glacial areas tend to be found mainly within the Antarctic Circle in places such as Antarctica. They are also found in Greenland. **a** The coast of Greenland, however, is dominated by a periglacial area. Periglacial areas are found within or near to the Arctic Circle, for example northern Russia, Alaska and North America. **b** Alpine areas tend to be outliers, distributed randomly and unevenly. **c**

Cold environments can be found in a range of different types of location. Some are found in high latitudes where the main causes are the weak insolation and reduced amounts of daylight due to the tilt of the Earth. Others are located in areas of high altitude where temperatures naturally decrease. **d**

ⓔ 4/6 marks awarded. a b c The answer begins by accurately describing the distribution of cold environments but with little sense of analysis. **d** The second paragraph explains this distribution — an outline of causes of a phenomenon is one way to address the command 'analyse'. The answer is unbalanced — more descriptive than analytical — yet does manage to access the lower end of Level 2.

Question 3

In Oymyakon the average daily temperature in November is −37°C. By then the Siberian winter has really set in, but temperatures will continue to fall. Oymyakon's lowest average daily temperature is in January (−46°C), but its coldest temperature of all was recorded in February 1921, at −71.2°C. That temperature is the lowest experienced at any permanently inhabited place, making Oymyakon the coldest permanently inhabited place on Earth. The town is near a river at a traditional camping spot for migratory reindeer herders and became established in the 1930s. The residents of Oymyakon inevitably have to deal with an extreme physical environment.

The main civic buildings are all in the centre of town and receive winter heating (in the form of hot water) and electricity from a local power station. Many houses lie outside the range of the power station's hot water distribution system and so are heated by their own wood-burning stoves. Wood is plentiful in the surrounding forest. Few homes in Oymyakon enjoy modern conveniences. Water is collected from the nearby river as chunks of ice that are stacked outdoors and brought inside the house to melt when needed. The lack of running water also means the toilet is outside — a wooden shed over a hole in the ground.

Pupils attend school throughout the winter, although they do get a day off if the temperature falls below −56°C. There is no mobile phone reception in Oymyakon. Everybody in Oymyakon possesses boots and a hat made of animal fur. The boots are usually made from reindeer hide, which is light but warming. Hats come in a variety of animal furs, including fox, raccoon, sable and mink. Many people also wear fur coats. Anyone who spends more than an hour or so outside in winter runs the risk of getting frostbite on exposed areas of skin, usually their cheeks.

Oymyakon's shops stock basic foods in tins and packets, plus some fresh root vegetables, but most locals have other food sources: hunting, trapping, ice-fishing, reindeer breeding and horse breeding.

The diet relies heavily on meat as a source of protein and fat to provide energy for the cold winter days. Horse meat is a particular staple. The horses of the region are well adapted to life in the cold and live outside all winter, although they have to be rounded up every month or two to scrape away the ice that builds up on their backs. Some people also keep cows for milk and cream, but the cows spend the whole winter in barns.

Using the above information, assess how the people of Oymyakon have adapted to the environmental conditions. (6 marks)

@ Mark scheme:
■ Level 2 (4–6 marks):
　□ AO1: Demonstrates clear knowledge and understanding of concepts, processes, interactions and change.
　□ AO2: Applies knowledge and understanding to the novel situation, offering clear analysis and assessment drawn appropriately from the context provided. Connections and relationships between different aspects of study are evident, with clear relevance. Clear statement(s) of assessment.
■ Level 1 (1–3 marks):
　□ AO1: Demonstrates basic knowledge and understanding of concepts, processes, interactions and change.
　□ AO2: Applies limited knowledge and understanding to the novel situation, offering basic analysis and assessment drawn from the context provided. Connections and relationships between different aspects of study are basic, with limited relevance. Tentative or no statement(s) of assessment.

Student answer

The people of Oymyakon have adapted to the environmental conditions in a number of ways. Two of the most basic needs that a person has are to be warm and have fresh water. In terms of heating, although many of the civic buildings have heating from a small power station, for their homes many people make use of wood-burning stoves, using the wood from surrounding forests. a Water is collected from the nearby frozen river in blocks, stored outdoors, and melted when needed. a This will mean that running water is unlikely to be 'on tap' — literally. b Sanitation is basic, though with an outside toilet being the norm. b

Other important needs include clothing and food. Much personal clothing is made from the hide and fur of animals, in particular the reindeer, fox, raccoon, sable and mink. a Temperatures are cold throughout the year, and are bitterly cold in the winter. People have to ensure that they do not get frostbite by being out in the cold for long periods without protection. b Much of the food supply comes from animals raised in the area — horses and reindeer — and fish.

There is a high intake of meat for protein and fats for energy. Cows are kept in barns for milk and dairy products a, which are essential parts of people's diets.

Although people have adjusted to living in such an inhospitable cold environment, it is not something most of us would like, especially not being able to use a mobile phone. b

e **5/6 marks awarded.** The answer addresses the thrust of the question throughout — adaptation. A number of needs of people in such an inhospitable environment are addressed, with some quoting of evidence from the data a together with some additional commentary, which can be viewed as assessment b. The only weakness with this answer is that it does not seek to examine other wider factors, such as access by transport, or health issues, that could possibly be inferred from the data. However, most aspects of Level 2 have been met. Mid Level 2 awarded.

Question 4

To what extent are recent developments in cold environments sustainable?   (20 marks)

e **Mark scheme: See the generic mark scheme on pages 56–57.**

Student answer

Tourism in Antarctica is very controversial. On the one hand it can be argued that it is sustainable because the tourists are carefully monitored, in small groups and all their rubbish is removed and disposed of elsewhere, in order to make sure they don't damage the environment. a Also, the majority of animals such as the Adelie penguin don't seem to be affected by human presence. However, the summer tourist season does interfere with the animals' breeding and migration patterns, and there is an increased risk of air and water pollution from the boats, which could even cause oil spills. Also, intrusive species such as Mediterranean molluscs are brought in on the cruise ships' hulls. There is also evidence that human footsteps could damage fragile mosses that exist there and these will take years to recover. b

There is also evidence that resource development is not sustainable. An example is the tar sand extraction in Alberta, Canada, that has made huge scars on the landscape and damaged the natives' traditional way of life. The 'dirty oil' has got into the river systems and the fish suffer huge deformities. Even the people have experienced more cases of cancer. Researchers in 2014 confirmed high rates of cervical cancers and a rare bile duct cancer among First Nations (Indian) communities who fish from the Athabasca River and hunt off the land. c From both a natural and a human point of view, these recent developments cannot be regarded as sustainable. d

However, further north in Alaska there is some fishing that is entirely sustainable. They use a better way to fish than trawling nets and they contribute millions of eggs to the sea from artificial hatcheries to maintain the fish stocks, an activity that is clearly highly sustainable. e

In Antarctica there has been legislation to limit development in order to protect the fragile environment, such as the 1971 Antarctic Treaty and the 1991 Environmental Protocol. ⓘ They got 44 countries to agree to only use Antarctica for scientific development, which is more sustainable, and they also banned mineral extraction, which is generally unsustainable. ⓖ

ⓔ **19/20 marks awarded.** The answer refers to a wide range of recent economic developments within cold environments, with specific detail provided for most. ⓐⓑⓓⓔⓕⓖ There is a strong sense of addressing the theme of sustainability throughout. The student also ends the majority of paragraphs with explicit statements of assessment 'to what extent'. ⓒ Within the section on the Canadian tar sands, the student introduces some quite detailed information and links to health. Overall, the argument of the answer is structured, though perhaps the Antarctica sections could have been placed together. However, this is nit-picking and all elements of Level 4 have been addressed, although perhaps a rounded conclusion could have been provided? High Level 4 awarded.

Knowledge check answers

1 (a) Negative feedback: surface temperature of the oceans increases through climate change — leads to increased evaporation — more clouds in the atmosphere — increases the amount of solar radiation reflected — decreases sea surface temperature.

(b) Positive feedback: atmospheric temperatures increase due to climate change — sea ice melts — ocean water absorbs more solar radiation than ice — ocean temperatures warm — more sea ice melts.

2 There are a number of factors that determine transpiration rates:

- **Temperature:** transpiration rates increase as the temperature increases, especially during the growing season, when the air is warmer due to stronger sunlight and warmer air masses.
- **Relative humidity:** as the relative humidity of the air surrounding the plant rises, the transpiration rate falls.
- **Wind and air movement:** increased movement of the air around a plant will result in a higher transpiration rate.
- **Soil-moisture availability:** when moisture is lacking plants can begin to senesce (age prematurely, which can result in leaf loss) and hence transpire less water.
- **Type of plant:** some plants that grow in arid regions, such as cacti and succulents, conserve precious water by transpiring less water than other plants.

3
- When rainfall intensity exceeds the infiltration capacity of the soil.
- When saturation of the soil takes place and any excess water must flow over the surface. This often takes place at the base of slopes.

4 Impermeable surfaces (roofing materials, concrete, paved driveways), roofs and roads are shaped to get rid of water quickly. Combined with a dense network of drains, this means that water gets to the river very quickly, reducing lag time and increasing discharge. Furthermore, some of these drains and culverts are inadequate, or become blocked by vegetation and/or litter, and hence flooding is quickly generated.

5 The impact of precipitation depends on its extent, its direction of travel, its intensity and its duration. Intense periods of rainfall tend to be of a shorter duration but they can have great impacts. Equally, longer duration events with lighter rainfall can cause flooding. The nature of precipitation can also have an effect — whether rain or snow. Snowmelt is a major cause of floods.

6 When volcanoes erupt they vent the gas to the atmosphere and cover the land with fresh silicate rock to begin the cycle again. At present, volcanoes emit between 130 and 380 million metric tonnes of carbon dioxide per year. For comparison, humans emit about 30 billion tonnes of carbon dioxide per year — 100 to 300 times more than volcanoes — by burning fossil fuels.

7 Carbon dioxide can be removed from fuel exhaust gases, such as from power stations. This carbon can then be stored in underground reservoirs, aquifers and even ageing oil fields. Carbon dioxide can be injected into depleted oil and gas reservoirs and other similar geological features.

8 Plants fix carbon dioxide from the atmosphere and the soils below the plants are rich in carbon from plant litter. High water tables in wetlands and peatlands create low oxygen conditions, which reduce rates of litter decomposition. Consequently, plant litter accumulates in the wet soils at a rate that exceeds losses from decomposition. Peat soils are highly organic, made up of predominantly of preserved plant litter, and are an important soil carbon store.

9 The EU has the European Union Emission Trading Scheme (EUETS). This is a mechanism that sets limits (caps) on the emission of a pollutant, but allows companies that are within the limit to sell credits (trade) to companies that need to pollute more. The power generation, steel, cement and other heavily polluting industries such as airlines are part of the scheme. Any industry with an account in the EU registry can buy or sell credits, whether they are a company covered by the EUETS or not. Trading can be done directly between buyers and sellers, through several organised exchanges or via the many intermediaries active in the carbon market. Australia has also adopted this scheme.

10 There can be storms of substantial size that can cause severe surface flooding. Having said this, most rain in desert areas has a low intensity, and in some coastal areas, such as Namibia and Peru, comes in the form of deposits from fogs.

11 Hadley cells refer to the circulation of air on either side of the thermal equator resulting from convection at that thermal equator and subsidence of air some latitudinal distance from it. At the surface air moves equatorwards, from high to low pressure, with an opposite movement of air at high altitude.

12 These are warm, dry winds that descend to the east of the Canadian Rockies, and similarly the Andes. Temperature rises of between 15°C and 20°C may be experienced, and this can cause rapid snowmelt and avalanche problems. Condensation and precipitation occur on the windward side of the mountain barrier and as the air begins to descend on the leeward side, it warms up rapidly. Because of the difference between the saturated and dry air, there is a net increase in temperature as the air crosses the mountains.

13 Mechanical weathering involves the breakdown of rocks into smaller fragments through physical processes, such as expansion and contraction due to temperature change. Chemical weathering

Knowledge check answers

involves the decay or decomposition of rock in situ by chemical processes only.

14 Star dunes have four or five arms extending radially from a central peak, and are formed where no one wind direction is dominant. They are up to 150 m high and between 1 km and 2 km across. If these coalesce they form into a serrated ridge called a draa dune, which can be up to 400 m high and tens of kilometres long.

15 Badlands are areas in semi-arid environments where soft and relatively impermeable rocks have been moulded by rapid runoff, which results from heavy but irregular rainfall. General features of badlands landscapes include:
- extensive development of gullies that erode headwards on hillsides, cutting into them and contributing to their collapse
- alluvial fans at the foot of steep slopes where smaller gullies emerge
- pipes, which are formed when water passes through surface cracks and carves out eroded passageways; pipes may also form caves when surface runoff is directed towards them
- natural arches, which are created by the erosion of a cave over a period of time.

An example of a badland landscape occurs in South Dakota, USA.

16 Wind: deflation hollows, regs, desert pavements, ventifacts, zeugen, yardang, barchans, star dunes, seif dunes. Water: wadis, badlands, alluvial fans, bahadas, salt pans. Desert slopes: inselbergs, mesas, buttes, pediments.

17 Overgrazing was not as a large a problem in the past because animals would be moved in response to rainfall, leading to a nomadic lifestyle. Today, however, people have a steady supply of food so they do not have to move about following rains. Nomadic farmers have become sedentary. Farmers use fences to keep their animals in one place, which causes overgrazing, and this causes trampling on the soil, which weakens its structure, and it blows away.

18 Fetch is the distance over which the wind has blown to produce waves. Tides are the periodic rise and fall in the level of the water in the oceans caused by the gravitational attraction of the sun and moon.

19 Strong currents can be present when the flow is away from the beach — towards the sea. These are known as rip currents. Flow velocities in rip currents often exceed the speed at which people can swim and, in combination with the large water depth, rip currents can easily drag unsuspecting swimmers out to sea.

20 In a flow there is a variation in the speed of movement, both laterally and with depth. Flows tend to have a higher water content. A slide is when materials in the regolith move with a large degree of uniformity, i.e. as a single unit. A slump has a rotational element to the movement.

21 Wave refraction is the tendency for waves to become parallel to the line of a coastline. The waves approaching a headland find the water shallows more quickly and movement is slowed down. However, waves in deeper water are unaffected and move more rapidly towards the bay. The line of the wave therefore begins to reflect the shape of the submarine contours. Erosion is also concentrated on the headland.

22 Wave-cut platforms continue to grow in width as waves break further and further away from the cliff face. This leads to a greater dissipation of wave energy such that eventually most waves will have little energy left to perform further erosion at the base of the cliff, and hence slowing down the growth of the platform. Eventually, therefore, platforms tend to act as cliff protectors. Some suggest that the maximum width of a wave-cut platform is 0.5 km, though this will vary according to wave height and strength.

23 The south and east of the UK is sinking by up to 0.8 mm per year and so experiencing faster rates of sea-level rise. Land in the north and west is rising at up to 2 mm per year so that relative sea-level change is lower. This change in land level in the UK is due to the loss of glacier mass at the end of the last ice age. The land in the north, where ice was lost, is rebounding, causing the south and east to dip downwards in a seesaw-like motion. Local isostatic change makes the impact of eustatic changes harder to predict.

24 Coastal erosion: cliffs, caves, arches, stacks, stumps, geos, blow holes. Coastal deposition: beaches, berms, spits, tombolos, bars, barrier beaches, dunes. Landforms created by sea-level change: rias, fjords, raised beaches, raised platforms, relic cliffs.

25 SMPs are managed by Coastal Groups, made up of local authorities and the Environment Agency. They are therefore a mix of local and national decision-making. This is important, as the money will come from both local and national sources, and the decisions made have to fit into a national strategy; what happens on one area of coastline may impact somewhere else, so there has to be some 'joined-up thinking'.

26 Accumulation is the net gain of a glacier or ice sheet. It includes any form of precipitation, but mainly snowfall, and avalanches from above. Ablation is the collective loss of water from an ice sheet or glacier. It can take the form of melting on the surface, internally or at the base. It also includes calving of blocks of ice where glaciers meet the sea, as well as evaporation and sublimation.

27 Glacial: areas covered by ice (glaciers, ice sheets and ice caps). Periglacial: areas fringing or in close proximity to ice, or with very cold temperatures. Alpine: very cold areas within mountainous regions.

28 Large valley glaciers in areas such as the Karakoram mountains in the Himalayas may be as long as 60 km and up to 2 km wide. Valley glaciers in the Alps can be several kilometres long and 1 km wide, whereas cirque glaciers may extend only a few hundred metres in all directions.

29 A glacier's surface velocity is highest near the centre and diminishes towards the sides at a fairly uniform rate, as friction against the rock walls will bring it close to zero. Velocity also tends to decrease with depth, especially in the lower parts of the glacier, nearer the bed. Again this is largely due to friction.

30 Surges are ice flows at relatively high speeds (over 10 km per year). They are caused by an instability that is induced when snowfall that collects in the accumulation zone is not transmitted downstream efficiently. Instead there is a prolonged storage of surplus snow and ice, which causes the glacier to grow in bulk to unstable proportions. Once a critical threshold is reached, and possibly triggered by another event such as an earthquake, the glacier begins to surge. The outcomes can be dangerous — a rapidly moving ice front, and if entering the sea, large icebergs resulting from calving.

31 There are a number of different forms of end moraine:

■ **Terminal moraine:** marks the furthest point of a glacier or ice sheet. It is found where the glacier snout ended and melted over a long period of time. It marks the point where ablation and accumulation were in balance for a long period of time (the snout was stationary).

■ **Push moraine:** formed where ice has re-advanced down the valley and pushed materials ahead of it and left a ridge of moraine.

■ **Recessional moraine:** formed where the ice has retreated up valley from the terminal moraine and left a new pile of unsorted debris during another period of standstill of the snout.

32 Continuous permafrost: covers the largest areas with mean air temperatures below −5°C. The ground can be frozen to depths of several hundred metres. Discontinuous permafrost: occurs over smaller areas with mean air temperatures between −5°C and −1.5°C. Its depth is much shallower, up to 35 m, and the surface tends to melt in summer. In such areas, rivers and lakes cause the permafrost to be absent around them due to their 'warming' effect. Sporadic permafrost: covers the smallest areas where mean air temperatures are between −1.5°C and 0°C. Permafrost occurs only in markedly cold spots.

33 Glacial erosion: corries, arêtes, glacial troughs, hanging valleys, truncated spurs, roches moutonnées. Glacial deposition: drumlins, erratics, moraines, till plains. Fluvioglacial: meltwater channels, kames, eskers, outwash plains Periglacial: patterned ground, ice wedges, pingos, blockfields, solifluction lobes, terracettes, thermokarst.

34 Firstly, the rate of soil thawing and carbon release seems to slow down after the initial burst of release. This could be because, as the soils warm up, new plants such as mosses grow on the surface and insulate the soil from further warming. In addition, the warmer climate means less snow in winter, which allows the winter cold to penetrate further into the soil, in effect storing up 'cold' in the soil to protect it from the warmth of the following summer.

35 The challenges are threefold:

■ The sheer size of the land areas involved across the 'top' of the world — northern Canada and Alaska, together with northern Asia.

■ The range of governments involved and the differing political principles of those governments — it would be difficult to imagine the governments of the USA and Russia, for example, coming to a consensus.

■ The practicalities and costs of bringing so many different people together — where, how and even what language in which to communicate.

Index

Note: **bold** page numbers indicate definitions of key terms.

Index

Index